AF595018

Energy Saving at Cities

Energy Saving at Cities

Editors

Francisco Manzano Agugliaro
Alberto Jesús Perea Moreno

MDPI • Basel • Beijing • Wuhan • Barcelona • Belgrade • Manchester • Tokyo • Cluj • Tianjin

Editors
Francisco Manzano Agugliaro
University of Almeria
Spain

Alberto Jesús Perea Moreno
University of Cordoba
Spain

Editorial Office
MDPI
St. Alban-Anlage 66
4052 Basel, Switzerland

This is a reprint of articles from the Special Issue published online in the open access journal *Energies* (ISSN 1996-1073) (available at: https://www.mdpi.com/journal/energies/special_issues/energy_saving_cities).

For citation purposes, cite each article independently as indicated on the article page online and as indicated below:

LastName, A.A.; LastName, B.B.; LastName, C.C. Article Title. *Journal Name* **Year**, *Article Number*, Page Range.

ISBN 978-3-03936-760-3 (Hbk)
ISBN 978-3-03936-761-0 (PDF)

© 2020 by the authors. Articles in this book are Open Access and distributed under the Creative Commons Attribution (CC BY) license, which allows users to download, copy and build upon published articles, as long as the author and publisher are properly credited, which ensures maximum dissemination and a wider impact of our publications.
The book as a whole is distributed by MDPI under the terms and conditions of the Creative Commons license CC BY-NC-ND.

Contents

About the Editors

Francisco Manzano Agugliaro, full professor at the Engineering Department in the University of Almeria (Spain), received his M.S. as Agricultural Engineer and Ph.D. in Geomatics at the University of Cordoba (Spain). He has published over 160 papers in JCR journals (https://orcid.org/0000-0002-0085-030X), H-index 26. His main interests are energy, sustainability, scientometrics, water, and engineering. He has served as a Supervisor of 25 Ph.D. Theses; Vice Dean of Engineering Faculty (2001–2004); Director of Central Research Services (2016–2019); Ph.D. Program Coordinator for Environmental Engineering (2000–2012), Greenhouse Technology, Industrial and Environmental Engineering (from 2010); and General Manager of Infrastructures (from 2019) at University of Almeria. Awards: Top reviewers in Cross-Field—September 2019 (Web of Science), 2019 Outstanding Reviewer Award (Energies), 2019 *Sustainability* Best Paper Awards.

Alberto Jesús Perea Moreno, associate professor at the Department of Applied Physics, Radiology, and Physical Medicine in the University of Cordoba (Spain), received his M.S. as Agricultural Engineer and Ph.D. in Geomatics at the University of Cordoba (Spain). He has published over 40 papers in JCR journals (http://orcid.org/0000-0002-3196-7033), H-index 12. His main interests are renewable energy, energy saving, biomass, sustainability and remote sensing. He served as the secretary of the Applied Physics Department (2017–2019) at University of Cordoba. Awards: 2019 *Sustainability* Best Paper Awards.

Preface to "Energy Saving at Cities"

The use of renewable energies and energy saving, and efficiency are the needs of a global society. According to the latest estimates, global energy demand could triple by 2050, and by then, 70% of the world's population will live in cities. Cities are currently responsible for 80% of greenhouse gas emissions, so they have a key role to play in shifting toward a sustainable energy future. Cities are threatened not only by overcrowding, but also by rising energy demand, obsolete infrastructure, volatile energy markets, and the effects of climate change.This book aims to advance the contribution of energy saving and the use of renewable energies in order to achieve more sustainable cities.

Francisco Manzano Agugliaro, Alberto Jesús Perea Moreno
Editors

Editorial

Energy Saving at Cities

Alberto-Jesus Perea-Moreno [1] and Francisco Manzano-Agugliaro [2,*]

[1] Department of Applied Physics, Radiology and Physical Medicine, University of Cordoba, Campus de Rabanales, 14071 Córdoba, Spain; aperea@uco.es
[2] Department of Engineering, University of Almeria, ceiA3, 04120 Almeria, Spain
* Correspondence: fmanzano@ual.es; Tel.: +34-950-015346; Fax: +34-950-015491

Received: 3 July 2020; Accepted: 17 July 2020; Published: 22 July 2020

Abstract: The use of renewable energies, energy saving, and efficiency are needs of global society. According to the latest estimates, global energy demand could triple by 2050 and, by then, 70% of the world's population will live in cities. Cities are currently responsible for 80% of greenhouse gas emissions, so they have a key role to play in shifting towards a sustainable energy future. Cities are threatened not only by overcrowding, but also by rising energy demand, obsolete infrastructure, volatile energy markets, and the effects of climate change. This Special Issue aims to advance the contribution of energy saving and the use of renewable energies in order to achieve more sustainable cities.

Keywords: energy saving; renewable energy; zero-energy buildings; energy efficiency; sustainability; sustainable transport; PV; energy saving in data processing centers

---

1. Introduction

Climate change is increasing due to the anthropogenic emission of greenhouse gases. The majority of these are due to the production and consumption of energy. The challenge for the future cities is the implementation of a mechanism that minimizes the need for injection of new energy resources in them, so that a high level of self-sufficiency should be achieved through the concept of circular economy, thus partially mitigating the impacts of climate change. Using solar energy today is considered to be one of the best solutions that can be installed in buildings to help with this issue.

This Special Issue aims to advance the contribution of energy saving and the use of renewable energies in order to achieve more sustainable cities. This Special Issue sought contributions spanning a broad range of topics related, but not limited to:

- Solar energy
- The use of rooftops for energy generation
- Energy conversion from urban biomass or residues
- Energy management for sewage water
- Bioclimatic architecture and green buildings
- Wind energy
- Cogeneration
- Public and private urban energy saving
- Policy for urban energy saving
- Smart meters
- Zero-energy buildings
- Legislations, regulations, and standards of energy

2. Publication Statistics

Details of the call for papers for this Special Issue, regarding the articles submitted being published or rejected, were: 10 articles submitted (100%), 3 articles rejected (33.3%), and 7 articles published (66.6%).

The regional distribution of authors by countries for the published articles is presented in Table 1, in which it is possible to observe 20 authors from six countries. Note that it is usual for an item to be signed by more than one author and for authors to collaborate with others from different affiliations. The mean number of authors per published manuscript was four authors.

Table 1. Authors' countries.

Country	Authors
Spain	6
Mexico	4
Palestine	4
China	3
Portugal	2
UEA	1
Total	20

3. Authors' Affiliations

This Special Issue's authors and their first affiliations are reflected in Table 2.

Table 2. Authors and affiliations.

Author	First Affiliation	Country	Reference
Azevedo, I.	University of Porto	Portugal	[1]
Leal, V.	University of Porto	Portugal	[1]
Abdallah, R.	An-Najah National University	Palestine	[2,3]
Juaidi, A.	An-Najah National University	Palestine	[2,3]
Assad, M.	An-Najah National University	Palestine	[2]
Salameh, T.	University of Sharjah	UAE	[2]
Manzano-Agugliaro, F.	University of Almeria	Spain	[2–4]
Abdel-Fattah, S.	An-Najah National University	Palestine	[3]
Perea-Moreno, A.J.	University of Cordoba	Spain	[4–6]
Alcalá, G.	Universidad Veracruzana	Mexico	[5,6]
Hernandez-Escobedo, Q.	Universidad Autónoma de Mexico (UNAM)	Mexico	[5,6]
Yi, P.	Sichuan University	China	[7]
Huang, F.	Sichuan University	China	[7]
Peng, J.	Sichuan University	China	[7]
Garrido, J.	Universidad Veracruzana	Mexico	[5]
Rueda-Martinez, F.	Universidad Veracruzana	Mexico	[5]
de la Cruz-Lovera, C.,	University of Cordoba	Spain	[4]
de la Cruz-Fernández, J.L.	University of Cordoba	Spain	[4]
G Montoya, F.	University of Almeria	Spain	[4]
Alcayde, A.	University of Almeria	Spain	[4]

4. Topics

Table 3 summarizes the research carried out by identifying the topics to which they belong, according to the proposed topics in the special issue. It was noted that two "Energy Saving at Cities" topics dominated the rest: "Solar Energy", and "Sustainability".

Table 3. Energy Saving at Cities.

Energy Saving at Cities	Number of Manuscripts	Reference
Solar Energy	2	[3,6]
The use of rooftops for energy generation	1	[3]
Energy conversion from urban biomass or residues	1	[2,3]
Wind energy	1	[5]
Public and private urban energy saving	1	[4]
Sustainability	2	[1,7]

Author Contributions: The authors all made equal contributions to this article. All authors have read and agreed to the published version of the manuscript.

Conflicts of Interest: There is no conflict of interest, the authors state.

References

1. Azevedo, I.; Leal, V. Factors That Contribute to changes in local or municipal GHG emissions: A framework derived from a systematic literature review. *Energies* **2020**, *13*, 3205. [CrossRef]
2. Abdallah, R.; Juaidi, A.; Assad, M.; Salameh, T.; Manzano-Agugliaro, F. Energy recovery from waste tires using pyrolysis: Palestine as case of study. *Energies* **2020**, *13*, 1817. [CrossRef]
3. Abdallah, R.; Juaidi, A.; Abdel-Fattah, S.; Manzano-Agugliaro, F. Estimating the optimum tilt angles for South-facing surfaces in Palestine. *Energies* **2020**, *13*, 623. [CrossRef]
4. de la Cruz-Lovera, C.; Perea-Moreno, A.J.; de la Cruz-Fernández, J.L.; Montoya, G.F.; Alcayde, A.; Manzano-Agugliaro, F. Analysis of research topics and scientific collaborations in energy saving using bibliometric techniques and community detection. *Energies* **2019**, *12*, 2030. [CrossRef]
5. Hernandez-Escobedo, Q.; Garrido, J.; Rueda-Martinez, F.; Alcalá, G.; Perea-Moreno, A.J. Wind power cogeneration to reduce peak electricity demand in Mexican states along the Gulf of Mexico. *Energies* **2019**, *12*, 2330. [CrossRef]
6. Perea-Moreno, A.J.; Alcalá, G.; Hernandez-Escobedo, Q. Seasonal wind energy characterization in the Gulf of Mexico. *Energies* **2020**, *13*, 93. [CrossRef]
7. Yi, P.; Huang, F.; Peng, J. A Rebalancing Strategy for the Imbalance problem in Bike-sharing systems. *Energies* **2019**, *12*, 2578. [CrossRef]

© 2020 by the authors. Licensee MDPI, Basel, Switzerland. This article is an open access article distributed under the terms and conditions of the Creative Commons Attribution (CC BY) license (http://creativecommons.org/licenses/by/4.0/).

Article

Analysis of Research Topics and Scientific Collaborations in Energy Saving Using Bibliometric Techniques and Community Detection

Carmen de la Cruz-Lovera [1,*], Alberto-Jesus Perea-Moreno [1], José Luis de la Cruz-Fernández [1], Francisco G. Montoya [2], Alfredo Alcayde [2] and Francisco Manzano-Agugliaro [2]

1 Departamento de Física Aplicada, Universidad de Córdoba, ceiA3, Campus de Rabanales, 14071 Córdoba, Spain; g12pemoa@uco.es (A.-J.P.-M.); fa1crfej@uco.es (J.L.d.l.C.-F.)

2 Department of Engineering, University of Almeria, ceiA3, 04120 Almeria, Spain; pagilm@ual.es (F.G.M.); aalcayde@ual.es (A.A.); fmanzano@ual.es (F.M.-A.)

* Correspondence: zr52crloc@uco.es; Tel.: +34-957-218-553

Received: 15 April 2019; Accepted: 24 May 2019; Published: 27 May 2019

Abstract: Concern about everything related to energy is increasingly latent in the world and therefore the use of energy saving concepts has been increasing over the past several years. The interest in the subject has allowed a conceptual evolution in the scientific community regarding the understanding of the adequate use of energy. The objective of this work is to determine the contribution made by international institutions to the specialized publications in the area of energy-saving from 1939 to 2018, using Scopus Database API Interface. The methodology followed in this research was to perform a bibliometric analysis of the whole scientific production indexed in Scopus. The world's scientific production has been analysed in the following domains: First the trend over time, types of publications and countries, second, the main subjects and keywords, third, main institutions and their main topics, and fourth, the main journals and proceedings that publish on this topic. Then, these data are presented using community detection algorithms and graph visualization software. With these techniques, it is possible to determine the main areas of research activity as well as to identify the structures of the collaboration network in the field of renewable energy. The results of the work show that the literature in this field have substantially increased during the last 10 years.

Keywords: energy-saving; energy conservation; energy utilization; energy efficiency; scientific collaboration

---

1. Introduction

The energy problem has become a global issue in a world where science and technology are undergoing a great evolution in a short period of time [1]. Consequently, the development of renewable energy and the use of energy saving concept has been increasing in last years, becoming a research focus around the world [2]. The growing interest in the subject has allowed a conceptual evolution in the scientific community regarding the understanding of the adequate use of energy.

While there is a close relation between the terms "energy saving" and "energy efficiency", there are certain distinctions that need to be clarified. The energetic saving entails a change in the habits of consumption. Sometimes it is enough to eliminate habits that waste energy. On the other hand, energy efficiency is the fact of minimizing the amount of energy needed to satisfy the demand without affecting its quality [3]; it supposes the substitution one machine to another that, with the same benefits, consumes less electricity. Therefore, this does not mean changes in consumption habits. The behaviour of the user remains the same, but less energy is consumed due to the consumption of energy to carry out the same service, which is lower. In Alcott, an extensive review of the literature on the subject of "sufficiency" can be found [4]. To achieve greater energy sustainability as much as possible, renewable energy and energy saving have been combined [5]. In this way, we will achieve a twofold purpose thanks to a combined strategy of energy saving and energy efficiency, without reducing the quality of life. In fact, it will be maintained, and even increased. These concepts are also making significant advances in the residential construction sectors [6]. Energy saving is not only linked to monetary savings [7], it also contributes to reducing greenhouse gas emissions [8], sustainable development and achieving a sustainable energy supply. However, the reduction of energy demand is not as easy to carry out as is thought and the actions that are imposed are in many cases insufficient to achieve real transformation [9].

Thus, it is necessary to emphasize that, sometimes, many improvements of the energetic efficiency do not reduce the consumption of energy in the amount predicted by the simple engineering models [10]. This is called a rebound effect, since the fact that energy services improve means that they are cheaper and, in turn, that the consumption of these services increases [11]. The Paris Agreement on climate change is an agreement within the framework of the United Nations Framework Convention [12] and energy consumption was the main point to deal with and the creation of an agreement worldwide. To achieve the objectives of this agreement, an outstanding change in the energy sector is necessary, the origin of at least two thirds of greenhouse gas emissions.

The transformation of the electricity sector, with renewable energy, has focused attention on a new debate on the design of the electricity market and electrical safety [13]. However traditional concerns about energy security have not disappeared. In addition, there are also pending issues related to energy access and affordability, global warming and CO_2 emissions. Therefore, the massive social concerns can be understood and lead to an increasing demand for ways to improve social and individual energy efficiency.

There remains an urgent demand for ways to improve the energy efficiency of society and people and a need to move increasingly to alternative, low-carbon or non-carbon energy systems. [14]. If we want to achieve sustainable development, energy efficiency becomes a key factor. It is a reality that only a few countries are engaged in research into renewable energy, representing between 70% and 80% of scientific production [15]. In general, countries are in the process of achieving, and in some cases exceeding, many of the objectives set in their commitments under the Paris Agreement. This is enough to reduce the expected increase in global CO_2 emissions related to energy, but it is not enough to limit the global warming to less than 2 °C. The increase in energy consumption, especially in HVAC systems (heating, ventilation and air conditioning), and CO_2 emissions have contributed to more energy policies focusing on the development of energy saving and efficiency strategies as an absolute priority [16]. An example of this is the European Energy Performance of Buildings Directive.

The major advanced economies: USA, the European Union, and Japan appear to be clearly on track to achieve their climate commitments, although it will be vital that these countries introduce other additional improvements in energy efficiency. A clear example of it would be the cement production industry, one of the most polluting. However, several mitigation measures are gaining importance in the recent years, and this industry is increasingly turning towards cleaner production. This is because new alternatives are used as raw materials and fuels. In existing industries, mathematical statistical models, simulation processes, optimization of procedures are everyday methods [17]. Natural gas and oil are going to continue to be the essence of the global energy system for years to come, as governments have indicated in terms of climate. Faced with this situation, the fossil fuel industry should consider the risks of a more direct transition. The interdependence between energy and water will become more pronounced in the upcoming years. In this sense, the management of the relationship between energy and water is crucial for the successful fulfilment of a series of development and climate objectives. It is evident that, to promote energy saving, it should be tried to create efficient energy systems and banish old ones. In fact, there is great potential for energy savings in the renovation of the housing stock in many cities, as is the case in Denmark, where 75% of the buildings were built before the eighties, when the first important demands were introduced for the energy performance of the building [18].

In this way, without altering the development of the technique, significant energy savings will be achieved. Nor should we ignore the human factor; the education and global awareness of citizens is important to achieve energy savings.

Recent studies show how construction technologies have been developed to improve comfort and energy savings in both conventional buildings using concrete or heavy structures, and also in timber buildings. An example of these technologies is the thermal insulation added to the facades and the improvement of the performance of the glazing or a district heating designed in a residential area. Nevertheless, although these technological advances can be implemented in concrete and wood buildings, the studies carried out highlight how timber construction is associated with thermal comfort and is perceived as innovative [19].

This is paradoxical because, after having expanded the use of gravel and concrete, the wood construction was increasingly neglected for a period of time. However, in recent years the Kyoto Protocol has pushed to limit CO_2 emissions in all productions. The use of sustainable materials is definitely being stimulated worldwide, as they can store carbon dioxide. The advantages of using sustainable materials have been confirmed by various scientific studies [20].

Buildings using sustainable materials are called green buildings. They reduce and optimize their energy and water consumption as much as possible and integrate into their environment, whether natural or urban, causing the least possible environmental impact. Smaller buildings tend to achieve these objectives satisfactorily. However, in larger green buildings it is much more difficult to meet the requirements of comfort, lighting, noise, use of space, or more hours of operation. It is feared that these innovative buildings will run the risk of repeating past mistakes, especially if they are too difficult to manage [21].

In this process of constructing eco-friendly buildings, it is fundamentally important to know the user's perception of indoor environmental quality and satisfaction in green buildings. Numerous studies show that employees perceive green spaces better than conventional buildings, but those surveyed are not able to differentiate between both types of buildings. However, there is an unequivocal correlation between ecological attitudes among employees of green buildings, showing them to be more satisfied and more prone to sustainable attitudes [22].

In order to achieve appropriate low-carbon rehabilitation interventions, users play a very important role. A sound learning process must be ensured for occupants, designers and building managers. The behavioural habits of citizens are themselves an essential component in reducing consumption patterns [23].

2. Materials and Methods

The aim of this work is to analyse all the world scientific publications of renewable energy indexed in the Scopus database in order to study the most important areas of investigation in energy saving, as well as to highlight the existing worldwide relationships between researchers and institutions. It is decided to select Elsevier's Scopus for this study as it is the largest database in the world. Thanks to the API interface, it is very easy to automate the search for literature related to the topic by selecting it by authors or by institutions. The two main scientific databases, Scopus and Web of Science pose the question of the comparability and consistency of the results of the different data sets [24]. With respect to journals reported in the two main scientific databases Scopus and Web of Science, a comparative study reveals that the volume of journals on the Web of Science is inferior to the number of journals in Scopus [25], and the relationships obtained in the results with both databanks for the items and the citation numbers by country, and for their rankings, are very robust ($R2 \approx 0.99$) [26]. This methodology was successfully used in several scientific fields, so only Scopus data has been used in this research [27].

2.1. Automated Data Collection Applying Scripts

The flowchart shown in Figure 1 explains how automatic data extraction would be carried out from Scopus. The process that is carried out by Research Network Bot (ResNetBot) [28] could be separated into the following stages: (1) Obtaining information on documents containing 'energy savings' as keywords or included in the title. We must notice that analyse the keywords in scientific articles is one of the most common research topic in different fields of engineering science. This information is represented graphically, where the size of the nodes is proportional to the h index and the lines between two nodes indicate that they are cited. (2) Obtaining information from authors of the works mentioned in step (1), which includes for each author the Scopus author identification number, affiliations, publications and dates, number of citations and h-index. (3) Processed to obtain the collaborative network of authors who have published articles that contain 'energy savings' as keywords or are included in the title. The information stored refers to the number of collaborations between co-authors and their affiliations, country and city. With all this information a graph will be created where the size of the nodes will indicate the scientific production according to the country and will be proportional to the H index and the distance between them will represent the collaboration with one another.

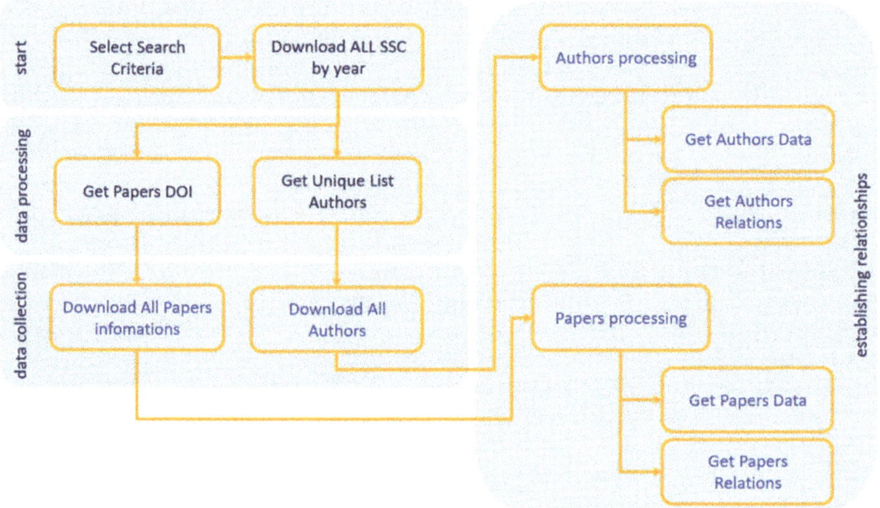

Figure 1. Flow diagram showing how the software obtains information from the Scopus database (ResNetBot).

2.2. Data Evaluation and Polishing of Texts

ResNetBot obtained data that was classified by domains and stored as a compendium of text files in JSON format., http://www.json.org. The Scopus API gives the option of requesting information through all type of details, so we programmed the bot to request all the information available in the files and choose the most convenient data for our work. As a result of the vast volume of data, some irregularities were observed during the work of checking the information. This, however, is a frequent and common problem in these large databases [29]. Phrases with a very similar meaning may sometimes appear written with slight changes. For this reason, it is very important to carry out a debugging process that allows us to obtain the appropriate information. The same expression may not be exactly the same written depending on whether we use capital letters in the first word, the second word, or both. We see an example of this, "Energy saving", "energy Saving", and "Energy Saving".

Therefore, some of the improvement criteria are applied from OpenRefine (http://openrefine.org), which contains some formulae such as clash of keys and neighbourhood algorithms to combine all expressions that correspond to the same idea but are not written in the same way. In order to classify data information and identify singular values, data sheets are finally used; this program has already been successfully carried out in former studies [30] and could be performed on key-words and author names.

Diagnostics and Data Viewing

The data collected by ResNetBot and further processed with OpenRefine was archived in a separate databank. Once this process has been completed, the latter information will be analysed using visualisation tools and statistics based on diagrams.

A graph is a set of vertices(elements) joined by edges. Graphically, the vertices are points and the edges are lines that join them, representing a relationship between two vertices. The visualization of graphs is a great help to determine the interrelation between the different parts. On the other hand, this representation makes it possible to incorporate specific values of the elements and their links. Over the last few years, powerful graphic representation programs have been created that enable the exhaustive analysis of the particularities of the graphics through the use of various functions, such as the reorganization of the vertices and the design of the vertices represented with different colours and sizes depending on their properties. One of the best known when using open source instruments is Gephi [31], which encompasses several statistical tools.

3. Results

3.1. Energy Saving Publication Relationship

In Figure 2, the documents obtained in the search described above have been represented all together. The graph represents the relationship between the documents, with each node representing a file. Of the total of 20,095 papers, in the outer halo formed by 12,402 individual papers are represented those that do not cite any other Energy Saving paper, this is 61.7%. Then there is another intermediate halo of 1919 papers that are linked to at least one other paper, but do not connect with the central core, which represent 9.6%. Finally, there is the central nucleus formed by 5774 nodes that are linked together, and that account for 28.7% of the total, is what could be defined as works that scientifically deepen in that field, because they are based on the previous ones to make some type of contribution.

In our research, there are a total of 70,951 keywords. As there is a large number of papers that are not part of the central core, we proceed to a refining of the 5774 core papers. 23,083 unique keywords were obtained, which after the refining process, 10,119 different keywords remain. After this process, the analysis of the documents that are related to each other is done, and they are represented according to the communities detected with the modularity algorithm of Gephi. Once this process is finished, Figure 3 is obtained. Of the 29 communities detected, the 15 main ones, with a percentage higher than 2%, have been represented in colour to distinguish them.

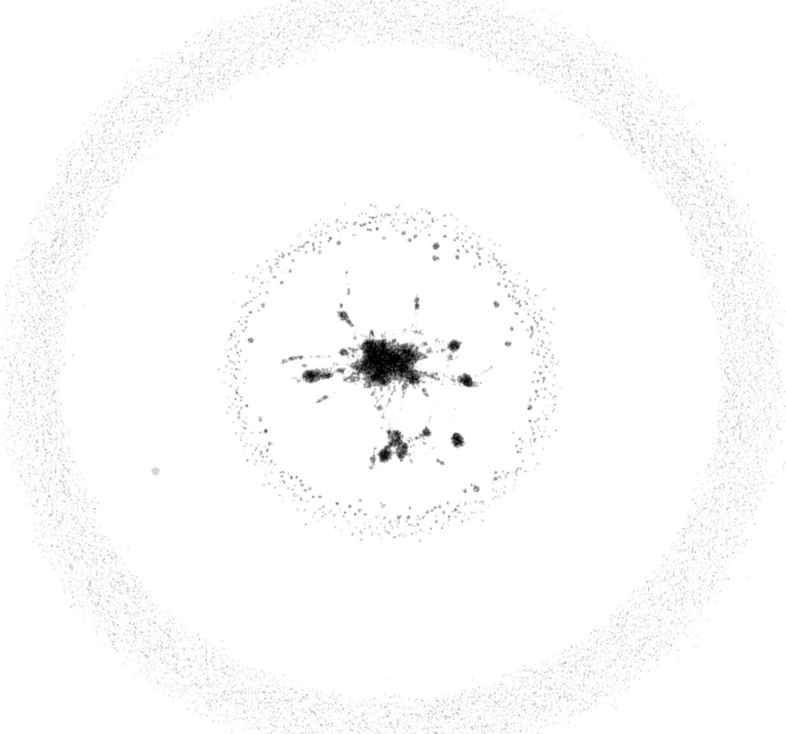

Figure 2. Display of the whole energy saving publication relationship indexed by Scopus.

This process has been necessary to reduce the excess of information that would impede the global vision we pursue.

Figure 4 shows the main 15 communities which have been named to facilitate the analysis. This table shows the main keywords of each community detected.

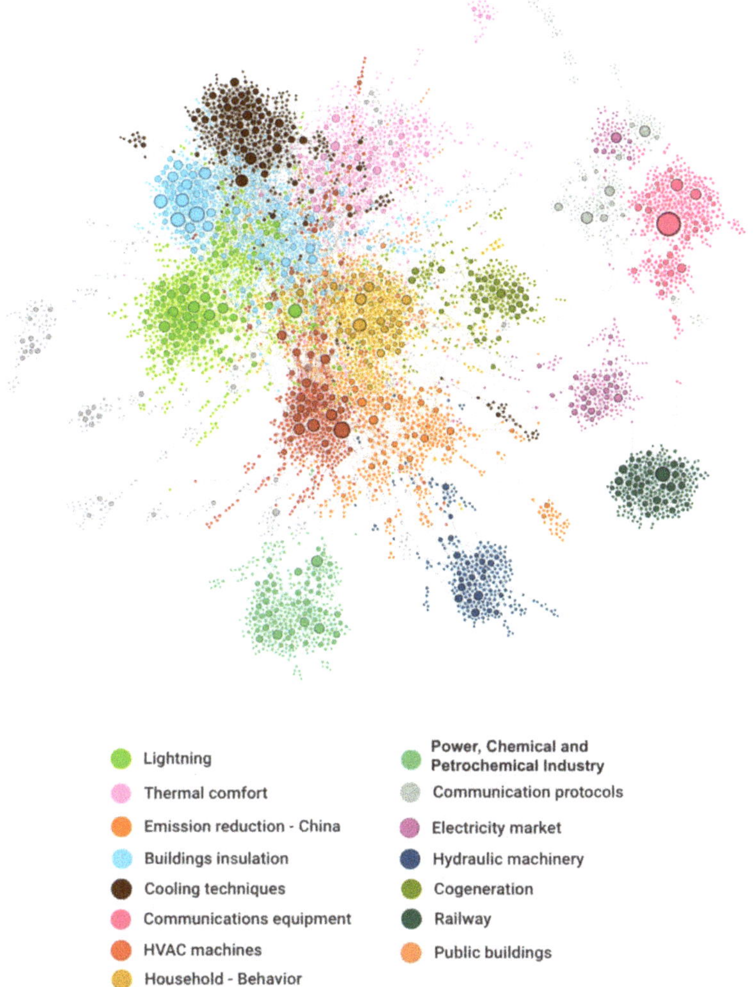

Figure 3. Display of the whole energy saving publication relationship indexed by Scopus.

Fifteen communities are detected within the large network of publications related to the term energy saving. In Figure 5, it can be seen that the area with the highest number of publications is lightning and thermal comfort, followed by emission reduction and building insulation. The high energy consumption that occurs in homes and businesses, together with the growing concern about the impact that energy consumption has on the environment, could show how important the search for facilities and systems that enable energy savings in these field is becoming. On the contrary, the railway and public buildings communities are in the last place of the ranking, with a percentage lower than 4%.

Lightning (9.44 %)	Thermal comfort (9.18 %)	Emission reduction (8.76%)	Buildings insulation (8.61%)	Cooling techniques (6.94 %)
Daylighting	Thermal comfort	China	Optimum insulation thickness	Phase change material
Energy efficiency	Energy efficiency	Energy saving potential	Buildings	Cool roof
Lighting control	Air conditioning	Energy intensity	Energy consumption	Energy efficiency
Daylight	HVAC	Energy saving and emission reduction	Residential buildings	Thermal comfort
Lighting	Indoor air quality	Emission reduction	Building envelope	Green roof
Lighting systems	Evaporative cooling	Energy consumption	Thermal comfort	Buildings
Energy consumption	Building energy saving	Energy saving device	Insulation	Building energy saving
Energy conservation	Natural ventilation	Data envelopment analysis	Thermal insulation	Passive cooling
Lighting system	Data center	Economic growth	Energy	Thermal energy storage
Thermal comfort	Liquid desiccant	CFD	Life cycle cost	Building envelope
Energy saving potential	Ventilation	Waste heat recovery	District heating	Urban heat island
Energy audit	Optimization	Energy-saving technology	Payback period	Solar reflectance
Building automation	Fuzzy logic	Sustainable development	Retrofit	PCM
LED	Office buildings	Iron and steel industry	Energy saving measures	Non-linear complex heat transfer
Optimization	Free cooling	Energy conservation	Insulation thickness	Finite element modelling
Energy saving system	Air conditioner			Sustainability

Communications equipment (6.29%)	HVAC machines (6.25%)	Household - Behavior (6.01%)	Power, Chemical and Petrochemical Industry (5.95%)	Communication protocols (5.02)
Heterogeneous networks	Emission reduction	Energy consumption	Distillation	IEEE 802.16e
LTE	Variable speed drive	Energy saving behaviour	Heat integration	Cloud computing
Cellular networks	Energy consumption	Behavior	Dividing wall column	Energy efficiency
Green communications	Thermal comfort	Household	Reactive distillation	Sleep mode
Sleep mode	Energy management	Residential buildings	Heat pump	Smartphone
Green networks	Efficiency	Residential sector	Process simulation	Energy consumption
Green networking	Air conditioning	Climate change	HIDiC	WiFi
Small cell	Energy	White certificates	Self-heat recuperation	WiMAX
Energy consumption	Optimization	Energy conservation	Extractive distillation	Cellular networks
Base station	HVAC system	Energy saving potential	Economics	Power control
Optimization	Compressed air systems	Energy audit	Process design	LTE
Green Internet	Energy conservation	Willingness to pay	Heat exchanger network	Mobile TV
Traffic Engineering	Energy audit	Occupant behavior	Simulation	offloading
Energy-saving management	Induction motors	Energy	Optimization	Mobile cloud computing
LTE-Advanced	Machine tool	Persuasive technology	Exergy analysis	Virtualization

Electricity market (4.75%)	Hydraulic machinery (4.47%)	Cogeneration (4%)	Railway (3.46 %)	Public buildings (2.63%)
Energy saving and emission reduction	Hydraulic excavator	Cogeneration	Optimization	Energy consumption
Electricity market	Excavator	Primary energy saving	Regenerative braking	Energy efficiency
Energy-saving generation dispatching	Accumulator	Trigeneration	Genetic algorithm	Activity recognition
Energy-saving and emission-reducing	Energy efficiency	Energy efficiency	Supercapacitor	School buildings
Energy-saving dispatch	Hybrid system	Optimization	Energy storage	HVAC
Smart grid	Energy recovery	Economic analysis	Energy-saving operation	Occupancy
Sleep mode	Hydraulic system	Operational planning	Electric railway	Energy management
Distribution network	Hybrid power	Emission reduction	Regenerative brake	Ventilation
Passive optical networks	Hydraulic transformer	Combined heat and power	Energy storage system	Energy audit
QoS	Load sensing	CHP	Speed profile	Wireless sensor networks
Energy efficiency	Control strategy	Energy saving rate	Simulation	Office buildings
EPON	Power matching	Solar energy	Efficiency	Thermostat
Unit commitment	Hydraulic accumulator	Supermarket	metro	Smart Home
Genetic algorithm	Construction machinery	Carbon dioxide emissions	Dynamic programming	Ambient intelligence
Power system	Mechanical engineering	Fuel cells	Traction power supply	IoT
				Context-awareness
				Schools

Figure 4. Communities detected for Energy Saving.

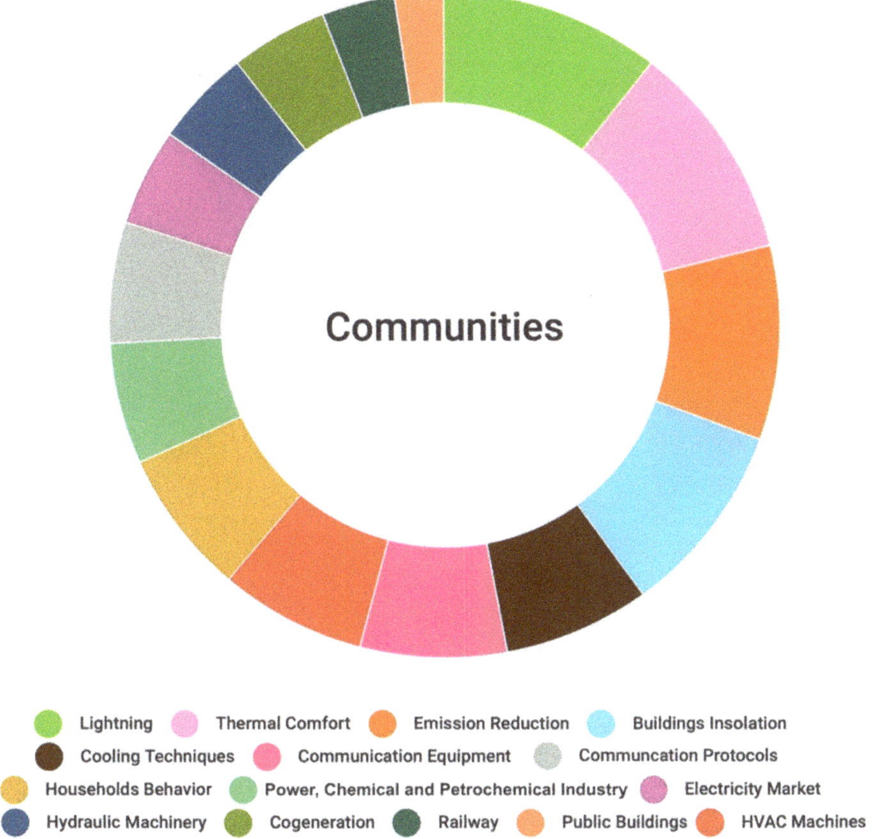

Figure 5. Communities are detected within the large network of publications related to the term energy saving.

3.2. *Authors, Affiliations and Countries in Energy Saving*

Regarding the authors and their relationship among them, it has been seen that in total there are 43,860 researchers. In Figure 6, the authors have been represented by colours according to the country of origin. This figure shows that the researchers that contribute most to the development of scientific production, come from academic and scientific institutions from China, which has the highest number of authors with more than half of the total number, followed by Italy (9.5%), the USA (8.9%), and Japan with 6.8%. Afterwards, with less percentage are Malaysia (4.77%) and Spain (2.2%), and finally a block of countries with a percentage below 2%. Null means that the Scopus register does not contain information on the country of the author.

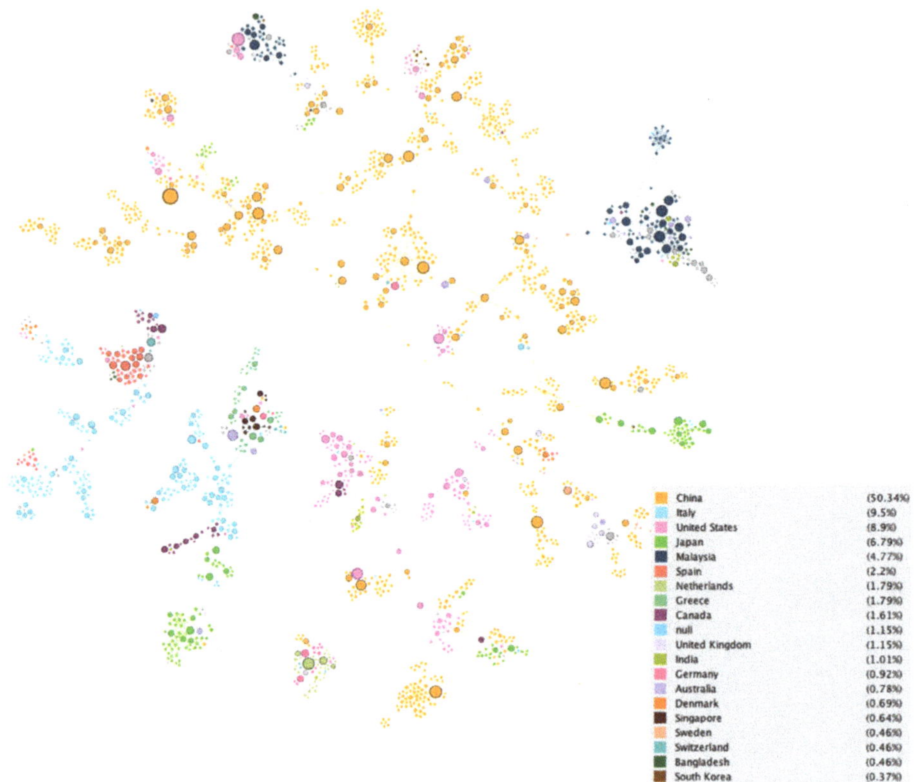

Figure 6. Collaborations between scholars from various countries. The size of the nodes is proportional to the H index.

From Figure 6 it can be drawn as conclusions that among Chinese researchers there is a tendency to collaborate among themselves, but also at the international level with American authors, Malaysians, and to a lesser extent, Japanese authors.

Whereas at European level, the relationship between its authors is with those authors belonging mainly to the European environment. These appear isolated, as can be observed in the lower left part of the graph, where it is extracted that the Italian researchers, the most productive, are mainly connected with the Spanish or with Greek authors among others, but also Japanese and Canadian.

Several layers are observed, being the most external (as explained above) that of the authors who have written about energy saving, but who do not collaborate with anyone in energy saving. Here are 2763 authors, which means a large minority. Then there are intermediate layers, where groups of collaborations are going from less to greater intensity. In this intermediate layer there are 41,097 authors, and it is observed that it is the most numerous. As a result of this international collaboration, many papers have been published.

The method is focused not only on the analysis of the scientific collaborations between institutions of each country, but also to determine which of these institutions are more involved in energy saving.

Researchers associated with Chinese institutions is shown in Figure 7, highlighting the number of people involved in energy savings in the following order: Tsinghua University (2.27%), North China Electric Power University (1.98%), and Tianjin University (1.91%).

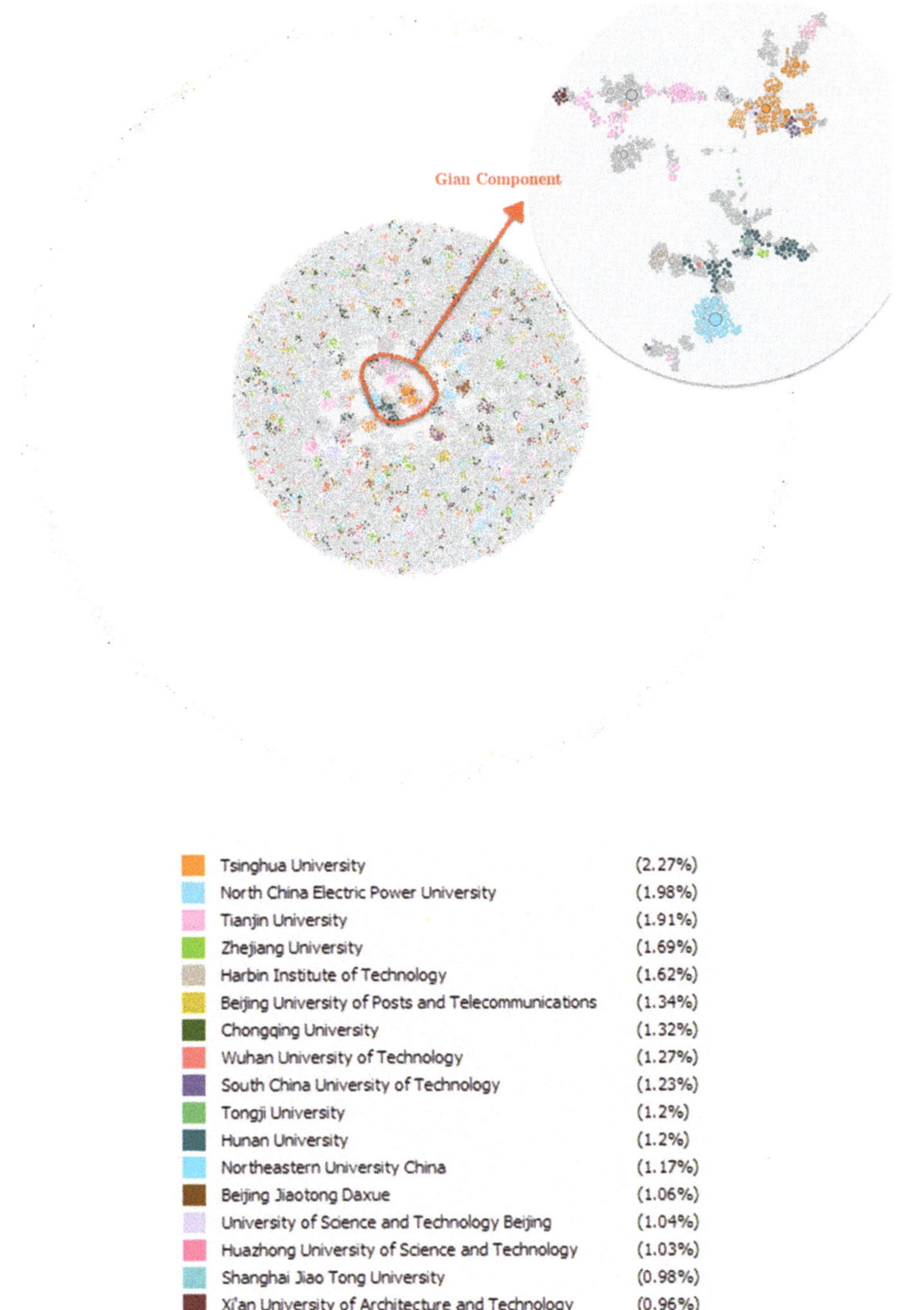

Figure 7. Connection of authors belonging to Chinese institutions.

In Figure 8 and Table 1 we can see the evolution of the publications developed by the most productive institutions worldwide in the last 10 years (2009–2018). On the bottom of the figure, the production of Ministry of Education of China can be seen, which is the most important institution registered. It does not surprise that, as a fact, it would englobe a lot of researching in the most populated country in the world. This institution represents the 1.64% of the total research (718 publications). Above it, the Tsinghua University represents the 1.59% (699 publications). This institution is one of the biggest universities in Asia with more than 40,000 students, 32,000 of which belong to masters or doctorate degree. After them, we have more institutions and as we can guess, all of them belong to China. It was a highlight surprise to notice all the research developed in this country and the high difference between China and other countries, at least in this search.

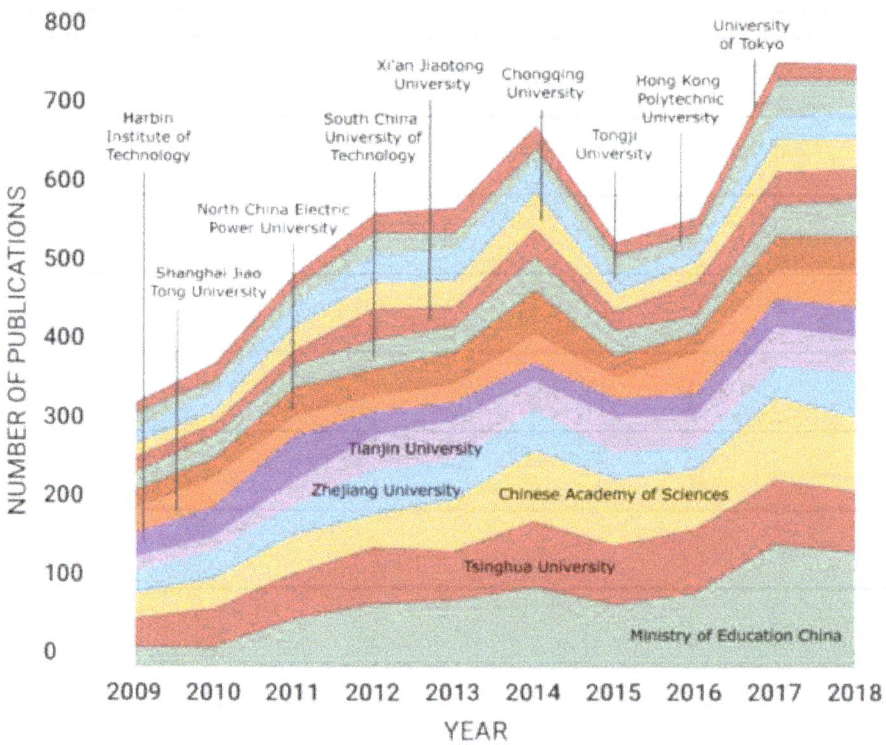

Figure 8. Ten most productive institutions during the period 2009–2018.

Table 1. Number of publications of the ten most productive institutions and universities between 2009–2018.

Institution	Total	%	2009	2010	2011	2012	2013	2014	2015	2016	2017	2018
Ministry of Education China	847	1.93	24	25	60	78	83	101	78	93	154	146
Tsinghua University	703	1.60	38	48	60	74	64	85	77	83	85	78
Chinese Academy of Sciences	685	1.56	30	40	46	41	64	88	82	73	103	92
Zhejiang University	443	1.01	29	35	38	56	50	53	34	31	42	59
Tianjin University	383	0.87	18	17	30	43	52	34	45	39	48	45
Harbin Institute of Technology	352	0.80	29	38	59	32	23	25	25	28	34	36
Shanghai Jiao Tong University	344	0.78	33	34	23	18	22	36	26	50	38	46
North China Electric Power University	376	0.86	22	28	38	36	41	57	29	26	44	45
South China University of Technology	337	0.77	22	27	26	36	32	40	31	23	38	45
Xi'an Jiaotong University	331	0.75	21	17	22	41	25	39	26	44	43	42
Chongqing University	303	0.69	18	16	28	33	34	45	22	25	39	37
Tongji University	308	0.70	16	27	32	35	40	37	23	21	32	36
Hong Kong Polytechnic University	260	0.59	21	11	20	27	20	17	23	14	46	39
University of Tokyo	235	0.54	14	22	18	27	33	30	20	22	23	20

3.3. *Typology of Literature, Languages of Publications and Distribution of Output in Subject Categories*

Table 2 shows the different types of publications during period 2009–2018. We can see how more than half of the total of publications of the search are articles (50.7%). Following them, the conference papers, with 43.1% of the total. After that, the quantity decreases to different types of documents as Reviews (2.3%), Book Chapter (1.6%), or Conference Revision (1.05%).

Table 2. Distribution of the searching documents according to the type of publications between 2009–2018.

Type	Publications	%
Article	18,510	50.67
Conference Paper	15,792	43.13
Reviews	848	2.32
Book Chapters	582	1.59
Conference Reviews	384	1.05
Article in Press	188	0.52
Short Survey	75	0.21
Note	70	0.19
Book	49	0.13
Editorial	32	0.09
Other	21	0.1

The vast majority of articles are published in international journals. Because of that reason, more than 87% of the work have been published in English (31,992), approximately 10% have been published in Chinese (3642) and only 3% includes other languages as Japanese, German, Russian, Polish, or Korean. In Table 3, shows the publications organized by language. Figure 9 shows the classification of publication results by subject, according Scopus.

Table 3. Most used languages in documents.

Language	Publications	Percentage (%)
English	31,992	87.31
Chinese	3642	9.94
Japanese	395	1.08
German	203	0.55
Russian	150	0.41
Polish	71	0.19
Korean	56	0.15
Spanish	54	0.15
French	48	0.13
Ukranian	29	0.08

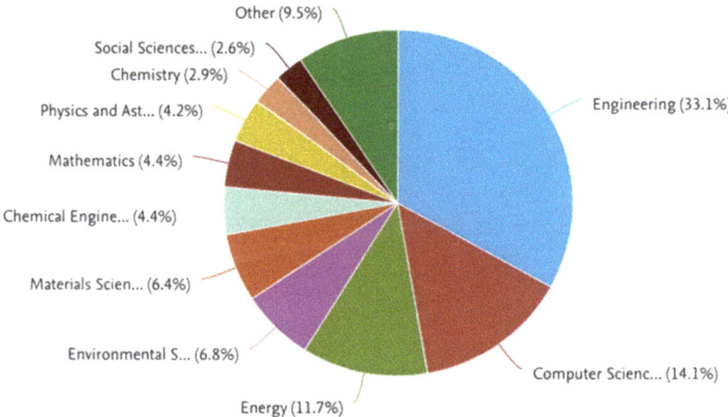

Figure 9. Classification of publication results by subject, according Scopus.

3.4. Literature Distribution by Area and Entity

Figure 10 shows the distribution of publications by country. On the map, red is associated with a high number of publications. Green indicates that there are less publications or grey means that there are no publications or a very limited number of them. It is easily observed how China is the country with more publications (18,309), followed by United States (3716), Japan, and Italy. Also highlight Germany, United Kingdom or Taiwan. This information leads us to the idea of how important the energy saving in the industrialised countries has become, as the richer states have also become involved in this issue by using harsh policies to achieve the proposed objectives. Figure 11 shows the 15 countries that have contributed the most in the last decade.

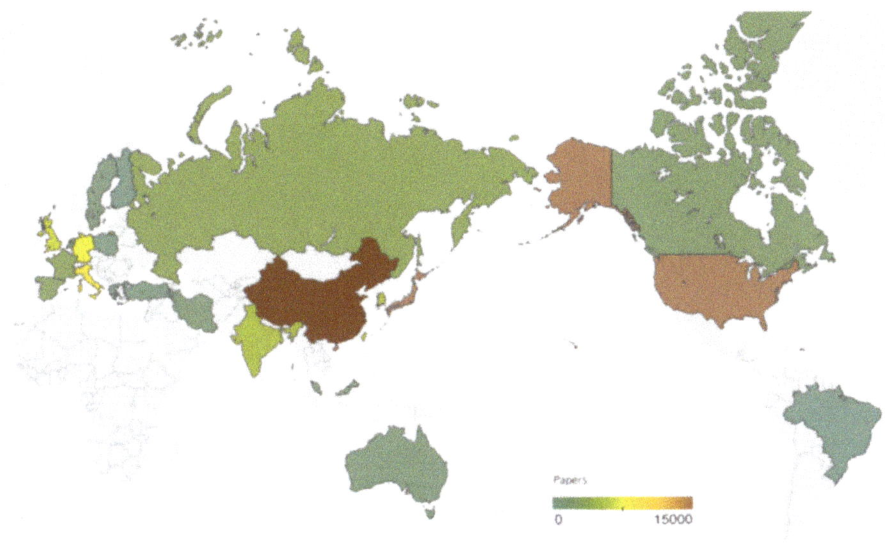

Figure 10. World Map with the 25 countries with a higher number of publications.

It should be taken into account the number of inhabitants of each country in order to properly analyse the results. It is not a big surprise to find out that China is the country with more publications as the population in China is the highest in the world (1,386,000,000 in 2017). This is followed by United States, also with a high population: 327,200,000 at 2018 and Japan with 126,800,000 in 2017. However, it is interesting to find countries such as Italy (60,590,000 in 2017) or Germany (82,790,000) with a lower population but still with a significant number of publications.

In Figure 12, the five keywords that appeared the most in our search are shown. 4/5 of these are related to energy, being the usual investigation and research of this field to create and find new ways of energy optimization and its use in new technologies. We can see how the term "Energy Conservation" started to be used in the beginning of the century and it has increased remarkably from 2013. In the second place, there is the term "Energy utilization" followed by "Energy Efficiency". After them "Energy saving" was very often used from 2008–2010, but it decreased as the researchers tend to focus their work in the conservation of the energy more than the saving of it.

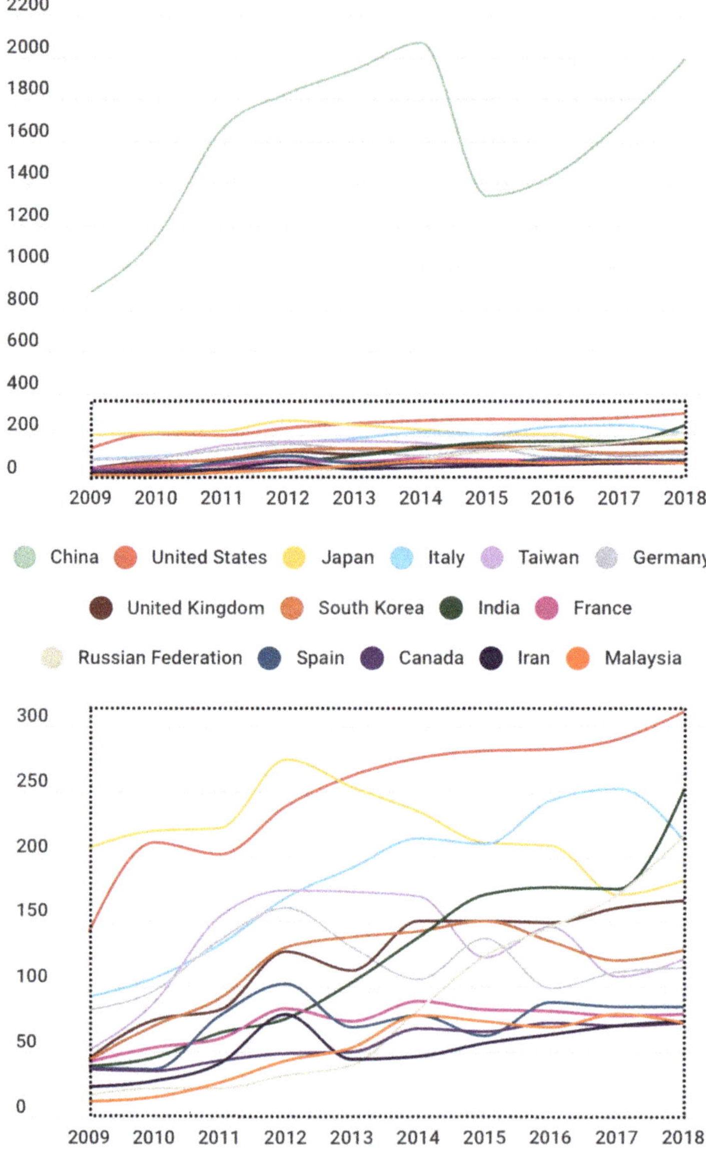

Figure 11. 15 countries that have contributed the most in the last decade.

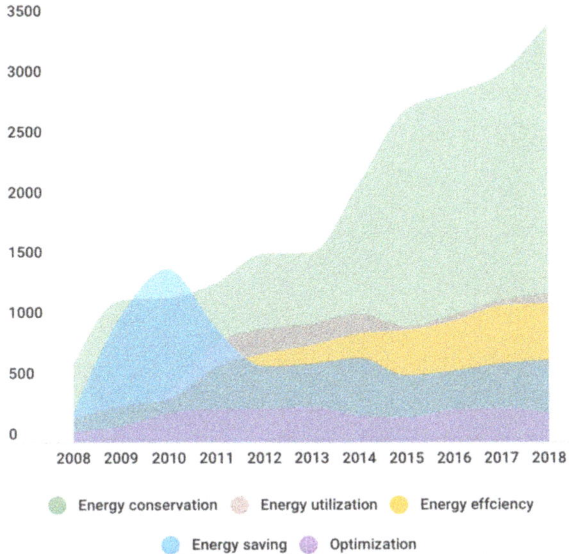

Figure 12. 5 most important Keywords and how often they were used during the last decade.

In Figure 13 and Table 4, the ranking of the six most important journals are shown. We have analysed not only the Scimago Journal Rank by Elsevier, but also the CiteScore. The metric of the latter calculates the citations of all papers submitted in the past few years preceding a title. It also includes Source Normalized Impact per Paper, SCImago Journal Rank, quotation and documents counting, and quotation rates. In the figure we have included also the H-Index as it has become very popular in the last decade and the reader can figure it out by the size of the circle according to each of the journals. We can see how "Renew. Sust. Energ. Rev." is the journal with higher not only SJR (3.04), but also CiteScore (10.54).

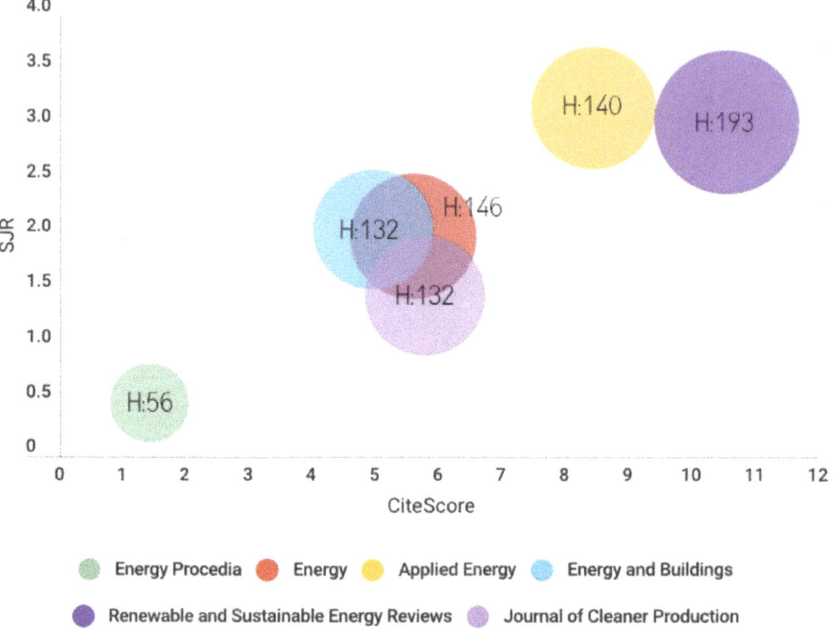

Figure 13. Classification of the first 6 journals. CiteScore is the Scopus ranking and SJR (Scimago Journal Rank) by Elsevier. The size of the circles indicates the H-index.

Table 4. Most important journals through the research.

Journals	Country	Quartile	JCR	H Index	Cite Score	SJR	SNIP
Energy Procedia	U. Kingdom	Q2	#N/A	56	1.44	0.49	0.799
Energy	U. Kingdom	Q1	4.968	146	5.6	1.999	1.923
Appl. Energy	U. Kingdom	Q1	7.900	140	8.44	3.162	2.765
Energ. Buildings	Netherlands	Q1	4.457	132	4.96	2.06	2.120
Renew. Sust. Energ. Rev.	United States	Q1	9.184	193	10.54	3.04	3.594
Journal of Cleaner Production	United States	Q1	5.651	132	5.79	1.467	2.194

The 21st century is the century of sustainability. This is why the building sector is investing a lot of resources in increasing energy efficiency, concerning both the materials and technologies used to make buildings and building techniques more energy efficient. In other words, it is not only a question of constructing more energy-efficient buildings, such as passivehause buildings and internship rehabilitation, but also of how to construct them, applying more efficient construction techniques, energy savings that must also be extended to the manufacture of construction materials and their transport to the construction site.

In Figure 14 we can see the most important authors related with the search through 2009 to 2018. The most important author is Kansha, who against all odds, is not from China, he is from Japan and has published a total of 57 papers related to the search, representing the 0.16% of the total of publications. He has a total of 1103 citations through all his work and a h-Index of 21. After him, Yang has a total of 47 works (0.13%) related to energy saving. All his work has a total of 8829 citations, getting a 45 H-Index. Other authors such as Li go after with 45 works (0.12%) and the Japanese Tsutsumi with 44 (0.12%).

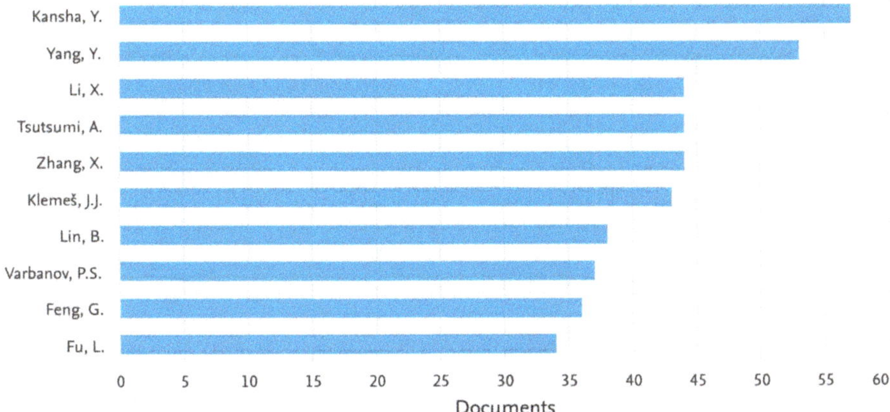

Figure 14. Authors who have published related with the search through the last 10 years.

4. Discussion

The scientific publications in relation to energy saving have had an exponential growth since the year 2000, from that date until now more than 90% of scientific production is condensed. This indicates that the last 20 years are those that have generated the interest of the States through their researchers and programs to support this research. For example, the research related to energy saving in communication and computation [32]. The works published in the topic of energy saving, are mainly in the form of scientific articles (50.7%), but there is an important number in the form of conference papers (43.1%), which indicates that today it is a novel field of research, in which this topic is addressed in many scientific meetings. It is observed that the books are a percentage below 1%, which means that it is not yet a very well established issue.

Regarding the language of publication, as expected, is dominated by English in more than 87.1%. However, when compared to other scientific fields, it is usual to be above 90%, which suggests that journals in particular countries address this issue in a significant way. As example the Dianli Xitong Zidonghua/Automation of Electric Power Systems, or Zhongguo Dianji Gongcheng Xuebao/Proceedings of the Chinese Society of Electrical Engineering. Such is the case of China, which contributes almost 10% of all publications in its own language. To a lower extent, German (e.g., the journal eb—Elektrische Bahnen) and Japanese (e.g., the journal IEEJ Transactions on Power and Energy from The Institute of Electrical Engineers of Japan), both around 2%, stand out. In any case, it is clear that China, as one of the world largest consumers of energy, has a great concern for energy saving [33], as was to be expected. This is also evident if it is observed that the main institutions that publish on this subject are also from China, Table 2. If the analysis focuses on the scientific fields where these works are assigned, Figure 9, it is remarkable that the first is not energy, but that the first place is occupied by engineering with 33.1%, showing that developments in this field are the largest part of the contributions, especially those related to energy conservation [34]. On the other hand, it is surprising that Computer Science plays such an important role in this field; this is justified by optimisation-related works [35]. Finally, it should be noted that one of the largest group is "others" with 9.2%. This should indicate the great diversity of scientific fields from which this subject is approached, being important the contribution to the energy saving that is studied from the greater number of points of view [36], which is what will allow to contribute to the global energy saving including geographical areas with serious energy shortages [37].

From the point of view of the networking of the works among themselves, this work shows a vast number of works published on the subject of energy saving, but the most important fact is that about 62% do not cite other work related to this subject. This implies great individualism, and scarce scientific connection in this field, which should make researchers reconsider, because it is important that research is based on previous studies to progress. However, it may also imply that they are researchers from other fields of knowledge who are becoming aware of or are including these items in their work, which is very encouraging. Within the works that are related to each other and are the core of scientific research in energy saving through the identification of communities have been detected the top 15. In the central positions, we find the scientific fields that might be said are most popular, see Figure 4: Lighting, thermal comfort, building insulation, cooling techniques, HVAC techniques, household-behaviour, or public buildings. Lighting has undergone a revolution since the emergence of solid-state light sources. The efficiency of solid-state technology provides, in a multitude of applications, energy savings, and environmental benefits [38]. Above all energy saving based on lighting has been studied for residential issues [39], public buildings, schools [40], or shopping malls [41]. Another one of the great challenges of energy saving is thermal comfort. Thermal comfort implies a good indoor climate that is important for the success of a building [42], not only by making its occupants feel comfortable, but also by deciding their energy consumption and thus influencing its sustainability [43]. On the other hand, persons have a natural trend to adapting to the conditions changing in their surroundings. This is especially the case with the scientific communities found around the central core of energy saving: building insulation, cooling techniques or HVAC techniques. Therefore, saving energy from households is one of the priorities of most countries, such us China, Italy, USA, or Japan. Those that occupy external positions are considered less integrated with the others, still have great possibilities of integration with the core and are generally the smallest. This is the way to find it: Distillation, communication protocols, electricity market, communication equipment, or railway. Related to these scientific communities it should be mentioned that distillation is still the most popular separation technology of chemical plants, and it uses around half of the operational costs, which shows the major concern of the researchers in applying here the energy saving. The energy savings in the communication protocols has been developed mainly using mobile telephony, where the terminals have a limited battery capacity. The third scientific community found is that related to the electricity market, being this one of variable tariffs, makes energy savings in certain time slots is of special interest. Regarding communication equipment, given the rising digital level of the developed countries, there is a growing concern about the data centres, especially in two respects, cooling and IT load. The amount of world energy consumption has gone from 0.5% in 2000 to 1% in 2005. With respect to the railway, the electric railway system is considered the most environmentally friendly means of locomotion. This is due to two important points: Firstly, the low resistance to running and secondly, the regeneration of energy during braking.

5. Conclusions

This study presents a wide range of data on the international contribution to scientific knowledge in the field of energy savings during the period from 1939 to 2018. The newness of this work is the application of community detection algorithms which identify scientific communities in the world that develop their work in fields related to energy saving. They also detect the collaboration preferences of their authors at an international level or even at an institutional level as in the case of China. It should be noted that many of the published papers have been the result of this international collaboration.

This methodology makes it possible to understand two complex systems: the network of publications based on quotations and the network of scientific collaborations based on the nationality of the co-authors.

Research Network Bot was used to search for publications indexed by Scopus containing 'energy savings', in the title or keywords. The information obtained was used for the analysis and graphical representation of these networks of scientific collaboration of researchers working in this field.

49,539 documents have been found, from which 55.5% are articles and 38.5% are conference papers. In this period there has been an exponential increase in publications, highlighting the categories of Engineering, Energy, and Computer science.

In light of the data obtained, Chinese institutions have provided an important contribution to the field. It is noted that, logically, the compound terms related with energy are the most used keywords, being energy conservation, energy utilization, energy efficiency, energy saving, and fifthly, optimization, proved as the most popular terms. Different aspects of the publications are analysed, such as language, type of publication, scope, top contributors' countries, institutions, journals, authors, as well as the frequency of appearance of keywords. Geographically, the giants China and United States, followed by Japan and Italy are highlighted as the top contributors' countries. The most active categories within the field of Energy are Lighting with 9.44%, Thermal Comfort (9.18%), Emission reduction and Buildings Insulation, both of then representing more than 8.5%. These 4 fields represent areas where energy saving plays a very important role as buildings, responsible for more than 40% of the energy consumption. In order to reduce this, there should be a release of a kind of regulation or policy to control systems for the installation of lighting.

Therefore, the communities that are in the lead, refer to the state of comfort, but also to the latent concern for the waste emitted into the atmosphere and how these can directly affect our planet in the future. Policies actually adopted in developed countries are insufficient to ensure a significant reduction of the energy consumption. It forecasts that the current global demand for energy in buildings will double by 2050. Thus, it is necessary to implement, at the global level, effective saving policies [44]. The challenge is always to reduce the energy consumption without forgetting the comfort of the user and the requirements of the building. The results of the work show that during the last 10 years have substantially increased the number of publications in this field, which indicates that research in terms of energy saving is assumed as an important and relevant issue in the international scientific scene.

During the last decade China, the most productive country in "energy saving" has become the main leader in the growth of energy efficiency. Its GDP grew a total of 7.8% per year and the energy intensity fell a total of 18.2%. This outstanding result is related with how energy efficiency policies have become the center of the political landscape through these year [45]. Energy efficiency was first included in the five-year plans, in the mid-2000s. It was a first sign of China's acceleration towards greener policies. In addition, energy efficiency has played a role in number of key Chinese economic policies: The market liberalisation and privatisations of the 1990s, the quest for energy security, economic development, or the fight against climate change.

This shows that there is a correlation between the number of publications in this country and the application of energy policies, but we do not have data to extend this assertion to the rest of the countries of the world. The study of the implementation of energy policies in each country, the energy savings achieved and the relationship with its scientific production on the subject, both quantitatively and qualitatively, would be the subject of future work.

Author Contributions: J.L.d.l.C.-F. and C.d.l.C.-L. performed the literature review and article writing. A.A. and A.-J.P.-M. analysed the data using community detection methods and article writing. F.M.-A. and F.G.M.: Research idea, article writing, and formatting; they jointly contributed to the structure and aims of the manuscript, paper drafting, editing, and review. All authors have read and approved the final manuscript.

Funding: This research received no external funding.

Conflicts of Interest: The authors declare no conflict of interest.

References

1. Wang, S.J. Discuss on building energy saving and energy saving materials. *Adv. Mater. Res.* **2012**, *512–515*, 2736–2739. [CrossRef]
2. Wang, J.; Li, L. Sustainable energy development scenario forecasting and energy saving policy analysis of China. *Renew. Sustain. Energy Rev.* **2016**, *58*, 718–724. [CrossRef]
3. De la Cruz-Lovera, C.; Perea-Moreno, A.-J.; de la Cruz-Fernández, J.-L.; Alvarez-Bermejo, J.A.; Manzano-Agugliaro, F. Worldwide Research on Energy Efficiency and Sustainability in Public Buildings. *Sustainability* **2017**, *9*, 1294. [CrossRef]

4. Alcott, B. The sufficiency strategy: Would rich-world frugality lower environmental impact? *Ecol. Econ.* **2008**, *64*, 770–786. [CrossRef]
5. Montoya, F.G.; Peña-García, A.; Juaidi, A.; Manzano-Agugliaro, F. Indoor lighting techniques: An overview of evolution and new trends for energy saving. *Energy Build.* **2017**, *140*, 50–60. [CrossRef]
6. Papadopoulou, E.V. Energy efficiency and energy saving. In *Energy Management in Buildings Using Photovoltaics*; Springer: London, UK, 2012; pp. 11–20.
7. Perea Moreno, A.J.; Hernandez-Escobedo, Q.; Jesus Aguilera-Urena, M. Water supply security and clean energy: A project proposal for irrigators in the river torrox. *Tecnol. Cienc. Agua* **2017**, *8*, 151–158.
8. AlFaris, F.; Juaidi, A.; Manzano-Agugliaro, F. Intelligent homes' technologies to optimize the energy performance for the net zero energy home. *Energy Build.* **2017**, *153*, 262–274. [CrossRef]
9. Sorrell, S. Reducing energy demand: A review of issues, challenges and approaches. *Renew. Sustain. Energy Rev.* **2015**, *47*, 74–82. [CrossRef]
10. Behar, O.; Khellaf, A.; Mohammedi, K. A novel parabolic trough solar collector model—Validation with experimental data and comparison to engineering equation solver (EES). *Energy Convers. Manag.* **2015**, *106*, 268–281. [CrossRef]
11. Sorrell, S.; Dimitropoulos, J.; Sommerville, M. Empirical estimates of the direct rebound effect: A review. *Energy Policy* **2009**, *37*, 1356–1371. [CrossRef]
12. Yun, G.; Xiang, G.; Zhang, X. The 2 °C global temperature target and the evolution of the long-term goal of addressing climate change—From the United Nations framework convention on climate change to the Paris agreement. *Engineering* **2017**, *3*, 272–278.
13. Cucchiella, F.; D'Adamo, I.; Gastaldi, M. A profitability assessment of small-scale photovoltaic systems in an electricity market without subsidies. *Energy Convers. Manag.* **2016**, *129*, 62–74. [CrossRef]
14. Dovì, V.G.; Friedler, F.; Huisingh, D.; Klemeš, J.J. Cleaner energy for sustainable future. *J. Clean. Prod.* **2009**, *17*, 889–895. [CrossRef]
15. Alcayde, A.; Montoya, F.G.; Baños, R.; Perea-Moreno, A.-J.; Manzano-Agugliaro, F. Analysis of Research Topics and Scientific Collaborations in Renewable Energy Using Community Detection. *Sustainability* **2018**, *10*, 4510. [CrossRef]
16. Pérez-Lombard, L.; Ortiz, J.; Pout, C. A review on buildings energy consumption information. *Energy Build.* **2008**, *40*, 394–398. [CrossRef]
17. Mikulcic, H.; Klemeš, J.J.; Vujanović, M.; Urbaniec, K.; Duić, N. Reducing greenhouse gasses emissions by fostering the deployment of alternative raw materials and energy sources in the cleaner cement manufacturing process. *J. Clean. Prod.* **2016**, *136*, 119–132. [CrossRef]
18. Tommerup, H.; Svendsen, S. Energy savings in danish residential building stock. *Energy Build.* **2006**, *38*, 618–626. [CrossRef]
19. Caniato, M.; Bettarello, F.; Ferluga, A.; Marsich, L.; Schmid, C.; Fausti, P. Thermal and acoustic performance expectations on timber buildings. *Build. Acoust.* **2017**, *24*, 219–237. [CrossRef]
20. Asdrubali, F.; D'Alessandro, F.; Schiavoni, S. A review of unconventional sustainable building insulation materials. *Sustain. Mater. Technol.* **2015**, *4*, 1–17. [CrossRef]
21. Leaman, A.; Bordass, B. Are users more tolerant of "green" buildings? *Build. Res. Inf.* **2007**, *35*, 662–673. [CrossRef]
22. Sant'Anna, D.O.; Santos, P.H.D.; Vianna, N.S.; Romero, M.A. Indoor environmental quality perception and users' satisfaction of conventional and green buildings in Brazil. *Sustain. Cities Soc.* **2018**, *43*, 95–110. [CrossRef]
23. Gupta, R.; Chandiwala, S. Understanding occupants: Feedback techniques for large-scale low-carbon domestic refurbishments. *Build. Res. Inf.* **2010**, *38*, 530–548. [CrossRef]
24. Perea-Moreno, M.-A.; Samerón-Manzano, E.; Perea-Moreno, A.-J. Biomass as Renewable Energy: Worldwide Research Trends. *Sustainability* **2019**, *11*, 863. [CrossRef]
25. Mongeon, P.; Paul-Hus, A. The journal coverage of web of science and scopus: A comparative analysis. *Scientometrics* **2016**, *106*, 213–228. [CrossRef]
26. Archambault, E.; Campbell, D.; Gingras, Y.; Lariviere, V. Comparing bibliometric statistics obtained from the web of science and scopus. *J. Am. Soc. Inf. Sci. Technol.* **2009**, *60*, 1320–1326. [CrossRef]
27. Aznar-Sánchez, J.A.; Belmonte-Ureña, L.J.; Velasco-Muñoz, J.F.; Manzano-Agugliaro, F. Economic analysis of sustainable water use: A review of worldwide research. *J. Clean. Prod.* **2018**, *198*, 1120–1132. [CrossRef]

28. Montoya, F.G.; Alcayde, A.; Baños, R.; Manzano-Agugliaro, F. A fast method for identifying worldwide scientific collaborations using the Scopus database. *Telemat. Inform.* **2018**, *35*, 168–185. [CrossRef]
29. Valderrama-Zurián, J.-C.; Aguilar-Moya, R.; Melero-Fuentes, D.; Aleixandre-Benavent, R. A systematic analysis of duplicate records in Scopus. *J. Informetr.* **2015**, *9*, 570–576. [CrossRef]
30. Montoya, F.G.; Montoya, M.G.; Gómez, J.; Manzano-Agugliaro, F.; Alameda-Hernández, E. The research on energy in spain: A scientometric approach. *Renew. Sustain. Energy Rev.* **2014**, *29*, 173–183. [CrossRef]
31. Bastian, M.; Heymann, S.; Jacomy, M. Gephi: An open source software for exploring and manipulating networks. *ICWSM* **2009**, *8*, 361–362.
32. Intanagonwiwat, C.; Govindan, R.; Estrin, D. Directed diffusion: A scalable and robust communication paradigm for sensor networks. In Proceedings of the 6th Annual International Conference on Mobile Computing and Networking, Boston, MA, USA, 6–11 August 2000; ACM: New York, NY, USA, 2000; pp. 56–67.
33. Wang, G.; Wang, Y.; Zhao, T. Analysis of interactions among the barriers to energy saving in China. *Energy Policy* **2008**, *36*, 1879–1889. [CrossRef]
34. De la Cruz-Lovera, C.; Manzano-Agugliaro, F.; Salmerón-Manzano, E.; de la Cruz-Fernández, J.-L.; Perea-Moreno, A.-J. Date Seeds (*Phoenix dactylifera L.*) Valorization for Boilers in the Mediterranean Climate. *Sustainability* **2019**, *11*, 711. [CrossRef]
35. Banos, R.; Manzano-Agugliaro, F.; Montoya, F.; Gil, C.; Alcayde, A.; Gómez, J. Optimization methods applied to renewable and sustainable energy: A review. *Renew. Sustain. Energy Rev.* **2011**, *15*, 1753–1766. [CrossRef]
36. Schultz, P.W.; Nolan, J.M.; Cialdini, R.B.; Goldstein, N.J.; Griskevicius, V. The constructive, destructive, and recon-structive power of social norms. *Psychol. Sci.* **2007**, *18*, 429–434. [CrossRef]
37. Juaidi, A.; Montoya, F.G.; Ibrik, I.H.; Manzano-Agugliaro, F. An overview of renewable energy potential in Palestine. *Renew. Sustain. Energy Rev.* **2016**, *65*, 943–960. [CrossRef]
38. Schubert, E.F.; Kim, J.K. Solid-state light sources getting smart. *Science* **2005**, *308*, 1274–1278. [CrossRef]
39. AlFaris, F.; Juaidi, A.; Manzano-Agugliaro, F. Energy retrofit strategies for housing sector in the arid climate. *Energy Build.* **2016**, *131*, 158–171. [CrossRef]
40. AlFaris, F.; Juaidi, A.; Manzano-Agugliaro, F. Improvement of efficiency through an energy management program as a sustainable practice in schools. *J. Clean. Prod.* **2016**, *135*, 794–805. [CrossRef]
41. Juaidi, F.; AlFaris Montoya, F.G.; Manzano-Agugliaro, F. Energy benchmarking for shopping centers in gulf coast region. *Energy Policy* **2016**, *91*, 247–255. [CrossRef]
42. Manzano-Agugliaro, F.; Montoya, F.G.; Sabio-Ortega, A.; García-Cruz, A. Review of bioclimatic architecture strategies for achieving thermal comfort. *Renew. Sustain. Energy Rev.* **2015**, *49*, 736–755. [CrossRef]
43. Nicol, J.F.; Humphreys, M.A. Adaptive thermal comfort and sustainable thermal standards for buildings. *Energy Build.* **2002**, *34*, 563–572. [CrossRef]
44. Berardi, U. A cross-country comparison of the building energy consumptions and their trends. *Resour. Conserv. Recycl.* **2017**, *123*, 230–241. [CrossRef]
45. Voïta, T. The Power of China's Energy Efficiency Policies. In *Études de l'Ifri*; Ifri: Paris, France, 2018.

© 2019 by the authors. Licensee MDPI, Basel, Switzerland. This article is an open access article distributed under the terms and conditions of the Creative Commons Attribution (CC BY) license (http://creativecommons.org/licenses/by/4.0/).

Article

Factors That Contribute to Changes in Local or Municipal GHG Emissions: A Framework Derived from a Systematic Literature Review

Isabel Azevedo * and Vítor Leal

Institute of Science and Innovation in Mechanical and Industrial Engineering (INEGI), Faculty of Engineering of the University of Porto (FEUP), Rua Dr. Roberto Frias, 400, 4200-465 Porto, Portugal; vleal@fe.up.pt
* Correspondence: iazevedo@inegi.up.pt; Tel.: +351-229578710

Received: 15 April 2020; Accepted: 16 June 2020; Published: 19 June 2020

Abstract: The changes observed in a municipal or local energy system over a period of time is due a number of concurrent dynamics such as the local social and economic trends, higher-level policies (e.g., national), and the local policies. Thus, a thorough identification and characterization of the many factors of change is key for an adequate assessment of the effectiveness of policies adopted at the local level, as well as for future planning. Such thorough identification and characterization of the factors that are determinant for the evolution of the local energy system is the content of this paper. The identification is performed through a systematic literature review, followed by a synthesis process. The factors identified are grouped into the categories of local context, local socio-economic and cultural evolution, higher-level governance framework, and local climate change mitigation actions. They are represented through a set of measurable variables. The factors and respective variables can be used to improve the disaggregation of changes in the local energy system into individual causes, leading to a better assessment of the evolution over time.

Keywords: local climate policy; municipal authorities; municipal energy planning; local GHG emissions; multilevel governance

---

1. Introduction

Local actions are crucial for the achievement of global climate change mitigation goals. Both the physical characteristics of the local energy systems and the regulatory competences of local authorities prompt local level as an appropriate level for action [1]. Moreover, local actions are becoming more common, due to the increasingly active role of local authorities and to the recognition of their importance by policy makers at higher governance levels. Thus, there is an increasing need to assess their contribution towards climate change mitigation, and to the achievement of global targets.

However, given the universal scope of climate policy and the link between the local energy system and local context, it is currently very difficult to properly assess the actual contribution of local policy actions. Indeed, the local energy system is influenced by numerous factors, of which only some are related with actions promoted by local authorities. The changes in the economic activity or on population are two examples. The assessment difficulties are amplified by the fact that local policies and actions are implemented as a part of a complex multi-level governance framework.

Previous studies have already highlighted the importance of considering external factors when evaluating local actions. Within the studies reviewed in [2], several factors were identified as key in the link between observed changes and actual effects of local climate change mitigation actions. For instance, the evolution of local socio-economic features (structure of local economy, demographic and economic conditions of local population, etc.), prompted by economic, social, and technological changes, is one of the aspects that are commonly considered relevant for the observed variation in local GHG emissions.

However, within existing studies, a comprehensive analysis of all the relevant factors that influence the local energy system and the corresponding GHG emissions was not found.

Moreover, the analysis of empirical studies focusing on assessing the effects of local actions has emphasized this need to thoroughly identify all the factors that contribute to changes in local GHG emissions [3–5]. For instance, in [3], Millard-Ball performed a quantitative analysis on the effects of city climate policy, without considering any external factors; although some scattered evidence of climate plans effects was found, the findings are not sufficiently robust (which could be partly explained by the non-consideration of other relevant factors). The evidence for impact of local climate plans in the different sectors is not demonstrated with the empirical analysis, which does not provide a coherent result across the different models that were tested. Similarly, in [5], the conclusions drawn were also insufficient to prove the causal link between local actions and GHG emissions. Within the results from the linear regression analysis with panel data, the coefficient associated with the effect of local actions is not statistically significant for any of the countries that were included in the study. However, in the same analysis, the relevance of municipality specificities as well as of higher-level context (i.e., country) for the assessment of local GHG emissions and their evolution over time was confirmed (resulting in coefficients that are statistically significant). Both these studies show the need for a comprehensive consideration of the different factors that impact local GHG emissions to ensure a robust and meaningful assessment of the specific impact of local climate change mitigation actions.

Thus, the identification of the factors of change of local GHG emissions is a crucial step towards a more complete understanding of the observed changes within the local energy system and assessment of the actual effects of local climate change mitigation actions. Indeed, to develop a model that is able to adequately represent the evolution of the local energy system, identifying the links between the changes in local GHG emissions and the different explanatory factors, it is necessary to first identify all the factors that may have an impact on local GHG emissions (factors of change). Herein, factors of change refer to all features and events that lead to changes in the local energy system and respective GHG emissions. These include factors related with local climate change mitigation actions as well as external factors (i.e., factors that are not directly influenced by actions promoted by local authorities). For instance, local demographic evolution—size and structure of local population—may lead to significant changes in the local energy system and, consequently, on the respective GHG emissions.

This paper attempts to perform a systematic identification and characterization of the factors of change of local GHG emissions. This comprises the identification of all factors that may potentially influence local emissions, even if their specific effects in the local energy system are not quantifiable or even relevant (depending on the local system being studied). In order for this to be applicable in the future, it is necessary not only to identify the relevant factors of change, but also to characterize them with indicators. Therefore, that is also covered by this paper. The work is presented in five sections. Section 2 describes the methodological approach that was used to identify the factors of change of local GHG emissions and the review process, including the documents used as reference to this analytical framework, as well as the presentation of the final list of factors identified. Section 3 is dedicated to the presentation of the results, including the description of the identified factors of change and the listing of the respective quantifiable indicators. Section 4 presents a brief discussion of the obtained results, focusing on their potential contribution for improved planning and evaluation of local climate change mitigation actions. The paper ends with some final remarks regarding the results obtained (Section 5).

2. Methodology

Considering the extensive research that has already been done on this topic, a wide literature review was considered an adequate methodological approach to identify and characterize the factors of change of local GHG emissions. More specifically, the method used consisted in a systematic literature review, using content analysis—which allowed reviewing, assessing, and aggregating a body of literature utilizing pre-specified standardized techniques to all the documents reviewed [6].

The definition of the specifications for data collection and analysis beforehand and respective use as a guide for performing the process aimed to avoid bias in the review.

A systematic literature review process consists of these four sequential steps: (i) definition of the objective; (ii) definition of the specifications to be followed; (iii) the retrieval and assessment of the information; and (iv) synthesis of the results obtained. In what concerns the specifications defined a priori as guidelines for the review, these consist in three types of rules/criteria: (1) the eligibility criteria to retrieve the literature; (2) the rules to retrieve information from literature; and (3) the rules for the analysis and screening of the obtained information. Table 1 presents the details of these three types of criteria.

Table 1. Systematic literature review specifications.

Objective	Identify and Characterize Factors of Change of Local Energy Systems
Documents Selection	
Search expressions	"Energy system characterization" "Energy system modeling/modelling" "Energy system planning" "Energy system scenarios" "Energy policy evaluation" "Local climate change mitigation"
Type of documents	Peer-reviewed articles Reports and working papers from internationally reputed entities PhD theses
Type of content	Factors that lead to changes in GHG emissions Characterizing variables of those factors Information on the relation between factors of change and local GHG emissions

The selection of relevant research covered the following keywords: "Energy system characterization"; "Energy system modelling"; "Energy system planning"; "Energy system scenarios"; "Energy policy evaluation"; and "Local climate change mitigation". The keyword selection was based on the different contexts within which this discussion has been raised, including planning, policy evaluation, and energy management.

To guarantee the reliability of the results, the documents considered in the review were restricted to peer-reviewed articles published in scientific journals or conference journals, reports or working papers published by internationally reputed entities (as e.g., International Energy Agency and European Union), and PhD theses. The reviewed documents were published between 1999 and 2017. A list of the documents with a description of their main focus of research is presented in Table 2.

Table 2. Characterization of the documents used for the identification of the factors of change.

Code	Study	Authors	Year	Topic/Theme	Doc. Type	Ref.
1	"Evaluating the effects of policy innovations: Lessons from a systematic review of policies promoting low-carbon technology"	Auld, G., Mallet, A., Burlica, B., Nolan-Poupart, F., Slater, R.	2014	Policy Evaluation	JP	[7]
2	"Energy and emission monitoring for policy use—Trend analysis with reconstructed energy balances"	Boonekamp, P.G.M.	2004	Policy Evaluation	JP	[8]
3	"Improved methods to evaluate realized energy savings"	Boonekamp, P.G.M.	2005	Policy Evaluation	PhD	[9]

Table 2. *Cont.*

Code	Study	Authors	Year	Topic/Theme	Doc. Type	Ref.
4	"A survey of urban climate change experiments in 100 cities"	Broto, C.B., Bulkeley, H.	2013	Multilevel governance	JP	[10]
5	"Reporting Guidelines on Sustainable Energy Action Plan and Monitoring"	Covenant of Mayors (European Union)	2010	Policy Evaluation	PR	[11]
6	"Cities, Climate Change and Multilevel Governance"	Corfee-Morlot, J., Kamal-Chaoui, L., Robert, A., Teasdale, P.J.	2009	Multilevel governance	WP	[12]
7	"The use of physical indicators for the monitoring of energy intensity developments in the Netherlands, 1980–1995"	Farla, J.C.M., Blok, K.	2000	Energy use	JP	[13]
8	"Long-term energy scenarios: Bridging the gap between socio-economic storyline and energy modeling"	Fortes, P., Alvarenga, A., Seixas, J., Rodrigues, S.	2015	Energy planning	JP	[14]
9	"Integrated technological-economic modelling platform for energy and climate policy"	Fortes, P., Pereira, R., Pereira, A., Seixas J.	2014	Energy planning	JP	[15]
10	"Multilevel governance, networking cities, and the geography of climate-change mitigation: two Swedish examples"	Gustavsson, E., Elander, I., Lundmark, M.	2009	Multilevel governance	JP	[16]
11	"Evaluating the effectiveness of environmental policy: A analysis of conceptual and methodological issues"	Gysen, J, Bachus, K., Bruyninckx, H.	2002	Policy evaluation	CP	[17]
12	"Model for Analysis of Energy Demand (MAED-2)—Computer Manual Series N° 18"	International Atomic Energy Agency	2006	Energy use	TR	[18]
13	"Energy Efficiency Indicators: Essentials for Policy Making"	International Energy Agency	2014	Energy planning	PR	[19]
14	"An analytical method for the measurement of energy system sustainability in urban areas"	Jovanovic, M., Afgan, N., Bakic, V.	2010	Energy use	JP	[20]
15	"Regions at Risk: Comparisons of Threatened Environments"	Kasperson, J.X., Kasperson, R.E, Turner, B.L.	1995	Policy Evaluation	Bk	[21]
16	"Evaluating the effectiveness of urban energy conservation and GHG mitigation measures: The case of Xiamen city, China"	Lin, J., Cao, B., Cui, S., Wang, W., Bai, X.	2010	Policy evaluation	JP	[22]
17	"Do city climate plans reduce emissions?"	Millard-Ball, A.	2012	Policy evaluation	JP	[3]
18	"A comparative analysis of urban energy governance in four European cities"	Morlet, C., Keirstead, J.	2013	Multilevel governance	JP	[23]
19	"A qualitative evaluation of policy instruments used to improve energy performance of existing private dwellings in the Netherlands"	Murphy, L., Meijer, F., Visscher, H.	2012	Policy evaluation	JP	[24]

Table 2. *Cont.*

Code	Study	Authors	Year	Topic/Theme	Doc. Type	Ref.
20	"Determinants of greenhouse gas emissions from Swedish private consumption: Time-series and cross-sectional analyses"	Nassen, J.	2014	Energy use	JP	[25]
21	"Outcome indicators for the evaluation of energy policy instruments and technical change"	Neij, L., Astrand, K.	2006	Policy evaluation	JP	[26]
22	"Simplified evaluation method for energy efficiency in single-family houses using key quality parameters"	Praznik, M., Butala, V., Senegacnik, M.Z.	2013	Energy use	JP	[27]
23	"Agent-based modeling of energy technology adoption: Empirical integration of social, behavioral, economic, and environmental factors"	Rai, V., Robinson, S.A.	2015	Energy planning	JP	[28]
24	"Indicators of Energy Use and Carbon Emissions: Explaining the Energy Economy Link"	Schipper, L., Unander, F., Murtishaw, S., Ting, M.	2001	Energy use	JP	[29]
25	"The impact of the economic crisis and policy actions on GHG emissions from road transport in Spain"	Sobrino, N., Monzon, A.	2014	Policy evaluation	JP	[30]
26	"A paper trail of evaluation approaches to energy and climate policy interactions"	Spyridaki, N.A., Flamos, A.	2014	Policy evaluation	JP	[31]
27	"Moving from agenda to action: evaluating local climate change action plans"	Tang, Z., Brody, S.D., Quinn, C., Chang, L., Wei, T.	2010	Policy evaluation	JP	[32]
28	"Global change in local places: How scale matters"	Wilbanks, T.J., Kates, R.W.	1999	Multilevel governance	JP	[33]

Legend: JP, peer-reviewed journal article; WP, working paper; CP, conference paper; Bk, Book; PhD, PhD thesis; PR, policy report; TR, technical report.

The information collected from the documents comprised three types of relevant information: (1) factors of change of GHG emissions; (2) variables that may be helpful to characterize the factors of change identified; and (3) information on the relation between the different factors of change and local GHG emissions.

In regard to the identification of the factors of change, Figure 1 presents a schematic representation of the whole process. The analysis of these documents resulted in a list composed of 511 variables with an impact on GHG emissions. The full list, as well as the reference to the documents where each variable has been referred, is presented in Table A1 (Appendix A).

Then, to reduce the list to the essential, both the eligibility criteria (Box 2 in Figure 1) and the screening filters (Box 3 in Figure 1) were applied successively. The following paragraphs describe this process in detail.

The eligibility criteria were defined to guarantee the applicability of the framework to the assessment of local climate change mitigation actions. Firstly, it was assumed that the respective change must be associated with energy-related GHG emissions, excluding, e.g., the emissions from crops and livestock. Then, the associated effect on the energy system should also be on Scope 1 and 2 emissions—GHG emissions that can be attributed to each municipality (i.e., that local inhabitants can be made accountable for). For instance, with this condition, changes in aviation and international commerce emissions are excluded, since typically an airport serves an area much larger than the municipality. Finally, it was also defined that factors should either change over time and/or influence the effect of the other factors identified. As the goal is to decouple the observed changes in a specific local energy system into effects of the different factors of change, it makes sense that factors that are constant over time (even if determinant for local energy use) must be disregarded. For example, topography of the local territory can have significant implications for local energy use, but it cannot be considered a factor of change as it cannot lead to changes of a specific local energy system over time.

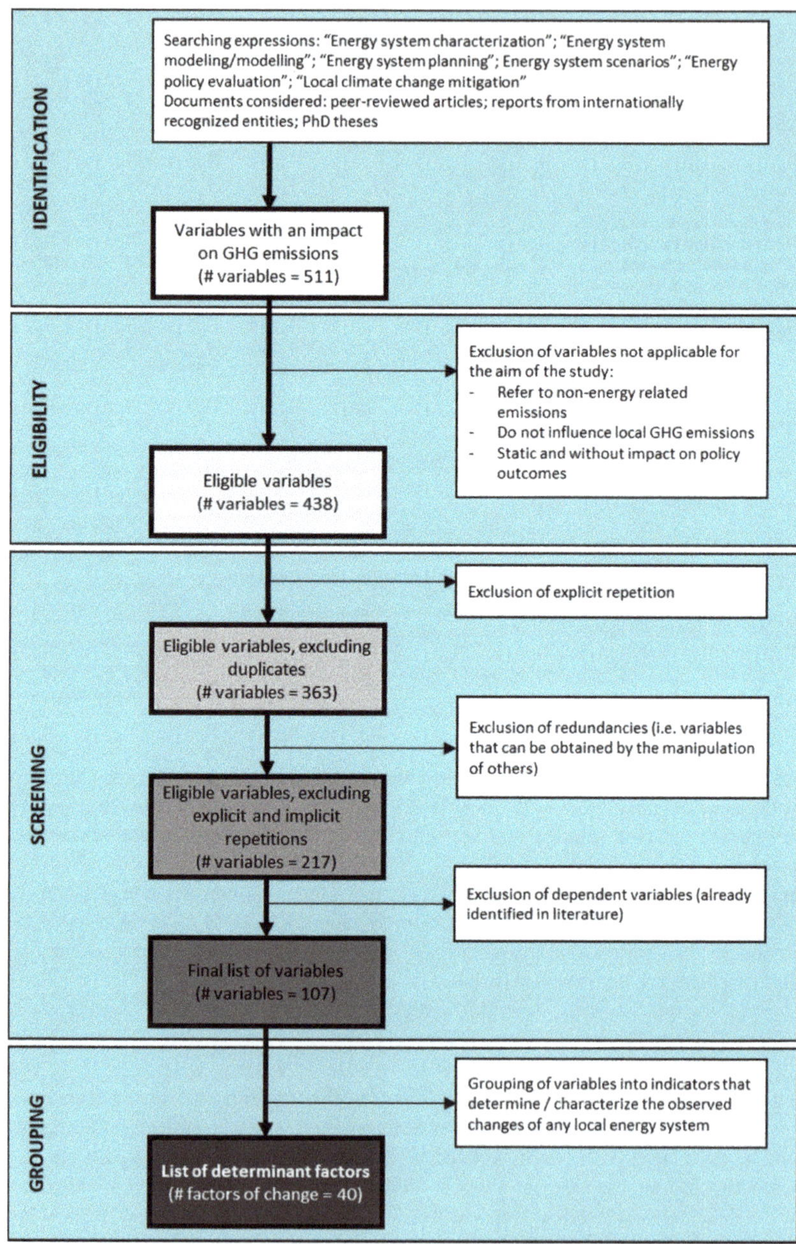

Figure 1. Schematic representation of the systematic literature review process.

The implementation of this step corresponded to the exclusion of variables that do not comply with at least one of the three eligibility criteria. The first eligibility criteria led to the exclusion of variables such as the "existence of strategies for landfill methane capture" and "changes in cattle production or fertilization strategies". Moreover, the constraint that implies that each factor must influence local GHG emissions led to the removal of variables such as "level of international trade" and "trade allowances between regions". Lastly, the criteria that exclude variables that do not change over time nor have an impact on policy outcomes implied the exclusion of factors "latitude", "topography", and others. Overall, with the application of the eligibility criteria, 73 variables were excluded from the list. The full list of the non-eligible variables is presented in the Appendix A (Table A2).

Furthermore, a screening process was defined and implemented to reduce the list of identified factors to the essential. This screening process is composed by three main steps that correspond to the application of three distinct filters: (1) explicit repetition; (2) implicit repetition; and (3) redundancy. The screening of the list of variables led to the exclusion of 331 variables. The full list of the excluded variables and the identification of the filters that led to their exclusion are presented in the Appendix A (Tables A3–A5).

The filter of explicit repetition refers to the removal of duplicates, i.e., factors that were identified by more than one document. For example, the number of inhabitants (or overall local population) and GDP were two of the variables that appeared more than once in the list.

Implicit repetition refers to factors that, even if not literally in duplicate, if withdrawn from the list do not imply an information loss. This includes factors that are a combination of other two or more factors that are also listed. For instance, average household area per capita can be considered as implicit repetition if both overall population and built area for residential purposes are already listed. This led to the exclusion of variables such as average household area per person (as total household area and overall population were also listed) and sectorial GDP (as GDP per subsector was also included).

Finally, the redundancy filter corresponds to the screening for dependent factors, whose relation has already been identified by literature. With this filter, for instance, average income per capita was excluded, considering that changes in private consumption (also linked to average income) would imply the exact same impact on local GHG emissions.

Once the variables with an impact on local GHG emissions have been identified and screened, a final list that is consistent with the goal of this work was obtained. The application of the eligibility criteria and the screening filters led to a final list of 107 variables that were grouped into 40 factors of change. This was done through an aggregation process, distinguishing policy-based factors from the ones that occur naturally, and factors related to actions implemented at local level from factors associated with governance actions at higher-levels. This grouping process is detailed in Table 3.

Table 3. Final list of variables and respective grouping into factors of change.

Factor of Change		**Variables Identified in the Literature Review**	
Statistical differences		3.002	Corrected statistics
Local inhabitants' characteristics	Educational level	17.004	Education
	Cultural diversity	15.021	Ethnic/religious views
	Willingness to act	18.006	Civic involvement (voting patterns)
		27.015	Concept of climate change (awareness of)
		27.016	Concept of GHG emissions (awareness of)
		27.017	Effects and impact of CC (awareness of)
Local administration characteristics	Competences	6.0005	Local authority's responsibilities and competences
	Qualified personnel	6.002	Existence of a dedicated institution/agency to deal with CC mitigation actions/plans
		6.003	Local authority's capacity and expertise on CC mitigation
	Economic situation	6.004	Local authority's ability to obtain funding
			Ratio between income and expenses

Table 3. *Cont.*

Factor of Change		Variables Identified in the Literature Review	
Climate variability		3.003	Climate (degree-days)
Demographic evolution	Population growth/change	15.002 15.003	Natural growth Migration
	Migration to urban areas	12.003	Share of urban population
	Age pyramid	14.003	Share of potential labor force
Social evolution	Household conditions	14.002 24.002 24.004 13.005 13.006 13.007 13.008	Persons per household Floor area per capita Heat per floor area Appliances ownership per capita DWH energy per capita Cooking energy per capita Lighting per floor area
	Commuting needs and transportation habits	13.010 12.023 13.011 27.009	Pkm Inverse of car ownership ratio Share of total pkm by mode Average commuting time
Economic evolution	Overall local economy	2.001	GDP growth
	Structure of local economy	8.004 8.005 8.006 8.007 12.012	GVA agriculture GVA services GVA transports GVA industry Distribution of sectors GVA by subsectors
	Economic situation of individuals	17.002 8.003	Employment Private consumption
Technology evolution	Market drivers/barriers	26.022 9.006 9.007	Technology market failures and other externalities related to electricity generation design Technical and costs evolution Availability and capacity limits (technology)
	Technical evolution	12.015 27.064	Average efficiencies of fuels for thermal uses per sector Energy intensity
	Technical evolution (Buildings)	14.010 24.005 13.017	Buildings insulation Intensity for different uses Energy per value-added (services)
	Technical evolution (Transports)	13.012 13.015	Energy per pkm by mode Energy per tkm by mode
	Technical evolution (Agriculture)	13.023	Energy per value-added by subsector
	Electricity generation and other energy transformation processes	3.010 3.011 3.012 3.013 3.014 3.015 3.016 3.017	Cogeneration end-users Fuel-mix end-users (actual energy use) Export (actual energy exports) Structure e-sectors (reference use refineries/gas supply) Cogeneration e-sectors Efficiency e-sectors (actual conversion efficiencies) Import (actual energy imports) Fuel-mix e-sectors
State and international governance framework	Existing policies	9.004 15.012 15.013	Energy and environmental policies International market National market
	Energy costs	18.012	Price of energy and relative share of any taxes or levies (energy costs)
	Synergies with local-level actions	6.001 6.006	Coherent multi-level policy framework (vertically and horizontally) Support from central governments to local policies

Table 3. *Cont.*

Factor of Change		Variables Identified in the Literature Review	
Actors involved	SEAP preparation	5.015	Decision body approving the plan
		5.003	Coordination and organizational structures created/assigned
		5.004	Staff capacity allocated
	SEAP implementation	10.001	Actors involved in local CC mitigation
		5.019	Origin of the action (local or other)
		5.020	Responsible body
Timeframe		5.014	Date of approval
		5.021	Implementation timeframe
Targets		5.001	Overall CO_2 reduction target
		5.002	Vision
Current implementation status		1.014	State of activity
		5.026	Newest inventory year
		5.031	Status implementation
		5.006	Overall estimated budget for the SEAP implementation
Estimated impact	Base year inventory	5.011	Final energy consumption per sector and carrier
		5.012	Energy supply sources
		5.013	CO_2 emissions per sector and carrier
	BAU projections	5.016	BAU projections
	Overall estimated impact	5.023	Energy savings
		5.024	Renewable energy production
		5.025	CO_2 reduction
	Mid-term inventory	5.028	Final energy consumption per sector and carrier
		5.029	Energy supply sources
		5.030	CO_2 emissions per sector and carrier
	Mid-term estimated impact	5.033	Renewable energy production from actions implemented so far
		5.034	CO_2 reduction from actions implemented so far
		5.009	Description of the impact of instruments
Budget	Overall	5.022	Estimated implementation costs
	Spent so far	5.032	Implementation cost spent so far
Financing sources		4.018	Funding
		5.007	Foreseen financing sources for the implementation
Actions characterization	Area of intervention	5.017	Area of intervention
	Technical measure	4.021	Action targeted sectors
		16.008	Measure/action
		16.009	Intensity of measure (degree of implementation)
	Instruments	4.012	Type of experiment
		4.014	Type of innovation
		1.017	Presence of flexibility mechanisms
		19.008	Policy theory associated with instruments applied
		5.018	Policy instrument
Other process characteristics		1.001	Agenda setting process
		27.046	Establish implementation priorities for action
		5.008	Monitoring process

Finally, to complete the theoretical framework, each individual factor was characterized with (preferably quantifiable) indicators. For most of the factors, the documents where they were extracted from also mentioned possible indicators. When more than one indicator was available, the choice fell back on the indicators whose data are more commonly available. When the documents did not mention possible indicators, these were searched for in national statistical agencies and policy reports from national or international governments. Within this step, there was an effort to choose indicators that use data that are commonly available for the municipality level, or equivalent scale, to avoid the need for downscaling.

3. Factors of Change of Local or Municipal GHG Emissions and Respective Characterization

This section presents the results obtained by the systematic literature review process described in the previous section. This includes discussion detailed description and an analysis of the final list of factors of change that were identified as well as the listing of the characterizing indicators for each factor of change.

To better understand the relation between the different factors of change and prompt a more fulfilling discussion, the 40 factors of change identified in the review were grouped into four distinct sets. Three of these groups correspond to three sources of change: local socio-economic and cultural changes; higher-level governance framework; and local climate change mitigation actions. The fourth group (herein referred to as local context) comprises variables that, despite not leading directly to changes of local GHG emissions, have influence on how the different sources of change impact local GHG emissions (e.g., climate awareness). A schematic representation of the relation between these sources of change and the observed changes in local GHG emissions, considering the local context, is presented in Figure 2.

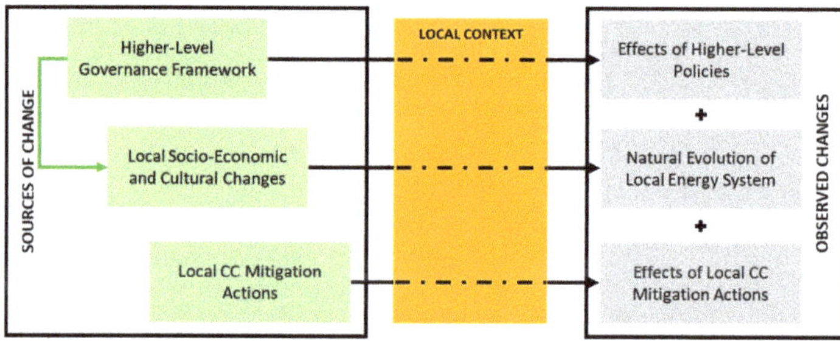

Figure 2. Framework of the sources of change and the observed changes in local GHG emissions.

3.1. Local Context

Local context refers to the local characteristics that, despite not leading directly to changes on the local energy system, influence the outcomes of both local and higher-level policies. On the one hand, this comprises the local inhabitants' characteristics, as education level, cultural diversity, and awareness towards the problematic of climate change. On the other hand, local context also includes characteristics referring to the local authority as an entity, including human and financial resources as well as legal competences. In general, the characteristics here referred as local context do not change significantly in the medium-term.

A complete list of the factors of change considered as local context is presented in Table 4, along with the respective characterizing indicators.

Table 4. Factors of change associated with local context with corresponding indicators.

Factor of Change		Characterizing (and Quantifiable) Indicator	Unit
Local inhabitants' characteristics	Education level	Distribution of local population per education level	Dimensionless (%)
	Cultural diversity	Valid permits by reason per 1000 inhabitants	# permits/1000 inhab.
	Willingness to act		
	Awareness	Share of waste that is separated previous to collection (recycling)	Dimensionless (%)
	Civic involvement	Number of environmental nonprofits per 1000 inhabitants	# groups/1000 inhab.

Table 4. *Cont.*

Factor of Change		Characterizing (and Quantifiable) Indicator	Unit
Local authority characteristics	Legal competences	Level of autonomy (High, Medium, Low)	Nominal variable
	Qualified personnel Dedicated institution/agency for CC mitigation	Y/N (or Number of people assigned to CC mitigation issues)	# people
	Previous experiences related with CC mitigation	Y/N	n.a.
	Economic situation		
	Ability to obtain funding	Global ranking of the municipalities' financial situation	Dimensionless
	Relation between income and expenses	Ratio between income and expenses	Dimensionless

A summarized description of each of the factors of change associated with local context is presented below.

3.1.1. Local Inhabitants' Characteristics

Education level refers to the average level of education of the local population, e.g., whether the majority of local citizens have completed high school, or even a bachelor degree. Here, it was assumed that the distribution of local population per education level could be a good indicator.

Cultural diversity refers to the heterogeneity of local population in terms of cultural background. This is mostly linked with the scale of immigration to the city, currently but also in the last decades. Culture is associated with daily habits and, thus, with specific patterns of energy use and reaction towards certain policy actions/incentives. One of the possibilities to assess this diversity could be through the number of residence permits by reason (or by country of origin).

Willingness to act refers to the level of motivation of the local citizens in having an active role towards climate change mitigation, which is highly associated with their level of awareness to the problem and their civic involvement in the community. For instance, the level of awareness of local population on environmental and climate change issues can have a huge influence on the outcome of any related policy independently on the level of implementation. Even if waste associated emissions are not considered within this work, it is assumed that the share of waste separation could be a good indicator of the local population awareness on climate change issues. Moreover, the degree of civic involvement may be assessed by the number and scale of environmental nonprofits that were created at the local level.

3.1.2. Local Authorities' Characteristics

Legal competences refers to the regulatory competences of local authorities regarding climate change mitigation, i.e., their scope of action. These may vary significantly from country to country and are determinant for the type of actions that are promoted at the local level. The level of autonomy of local authorities can be seen as a good indicator of their scope of action.

Qualified personnel corresponds to the technical competences of the municipal staff for developing and implementing actions targeting climate change mitigation. This can be assessed through the existence of a dedicated agency or municipal department to climate change mitigation, and from past experiences with these issues.

Economic situation translates the economic capacity of the local authority in financing and promoting actions towards climate change mitigation. This can be assessed by the relation between income and expenses of the local authority, which provides a characterization of their yearly profit or debt. Moreover, the ability to obtain funding, through private partnerships or public tenders, is also a relevant indicator.

3.2. Local Socio-Economic and Cultural Changes

Local socio-economic and cultural changes refer to the group of changes that result from local demographic, social, and economic changes, thus not directly caused by energy-related policies. In planning terms, this would correspond to the "business as usual" or "reference" scenario, i.e., the potential evolution of the system between two distinct moments in time, considering that no policies would be implemented in this timeframe. This comprises demographic, social, and economic changes, as well as climate variability.

The full list of factors of change associated with the local socio-economic and cultural changes that have been identified is presented in Table 5, along with the respective characterizing indicators.

Table 5. Factors of change associated with local socio-economic and cultural changes with corresponding indicators.

Factor of Change		Characterizing (and Quantifiable) Indicator	Unit
Climate variability		Heating Degree Days Cooling Degree Days	HDD CDD
Demographic evolution	Population growth	Change of overall population	# inhabitants
	Migration to urban areas	Share of rural population	Dimensionless (%)
	Age pyramid	Share of active population	Dimensionless (%)
Social evolution	Household conditions	Persons per household Floor area per person	Person/household m^2/cap
	Commuting needs	Average commuting needs	km/person.day
	Transportation habits	Car ownership	cars/cap
Economic evolution	Overall growth of local economy	Local GDP change	€
	Structure of local economy	GVA per sector (or number of employees)	€
	Economic situation of households: Employment Private consumption	Unemployment rate Private Purchase Power	Dimensionless (%) €/cap

3.2.1. Demographic Evolution

Demographic changes include the population growth (due to migration and natural growth) and also changes in the age pyramid and in the share of population living in urban and rural areas. All these variations imply changes in the local energy system—in the overall energy use as well as in the structure of this use. For example, energy use in urban areas differ from that in rural areas, due to the differences in heating options available (natural gas vs. traditional wood) and the different household conditions (housing type and size). A summarized description of each of the factors of change associated with demographic changes is presented below.

Population growth refers to the changes regarding the size of local population that occur within the studied timeframe. This is assessed through the number of local inhabitants and their evolution over time.

Migration to urban areas translates the level of urbanization of a municipality and its evolution over time. A change in the level of urbanization has an impact in the local energy system, in terms of preferred transportation modes, average distances traveled, and even in the type of housing. The changes in the level of urbanization can be assessed through the share of local population living in rural/urban areas.

Age pyramid refers to the distribution of the population according to their age. A change in this distribution implies changes in the local energy system, caused by the distinct energy service needs. The differences are more significant between active and inactive citizens. Thus, it is assumed that this factor of change can be assessed by the quantification of the share of active population.

3.2.2. Social Evolution

Social changes refer to the lifestyle changes of local inhabitants that occur naturally, not promoted by energy-related policies. These may include e.g., the increase of appliances ownership per capita and changes in the number of persons per household, in the residential sector, and a change in the average distance between home and work or school, in the transportation sector. Each of the factors of change associated with social evolution are described below.

Household conditions refer to lifestyle changes associated with the residential building. This includes changes in the average number of people per household (usually also linked with the evolution in families' typology) and changes in the average size of the household. These changes can imply significant changes in the energy use for space heating and cooling, lighting, and even cooking. Moreover, changes in appliances ownership must also be considered.

Commuting needs reflect the distance between home and work/school for the local inhabitants. A variation in this distance implies also a change in the daily transportation needs, and often even a change in the transportation modes that are used for commuting. It is assumed that this factor can be assessed through the average distance traveled per person per day, for commuting purposes.

Transportation habits refer to changes in the preferences for daily traveling, for both commuting and leisure purposes. These characteristics may change over time for multiple reasons—economic, social or other. It is assumed that a large share of these changes can be estimated based on the changes in car ownership of local population.

3.2.3. Economic Evolution

Economic evolution refers to the economic situation of local inhabitants and also to the evolution of local economic activities. The effect of these changes on local energy system may be affected, even if not in a very significant way, by local context. For instance, the increase in average income of local population leads to an increase in energy use; however, the growth rate may depend on the level of awareness of the local inhabitants towards climate change issues. The factors of change associated with the economic evolution at the local level are described below.

Overall growth of local economy refers to changes in the level of activity, i.e., the overall value of the local economy. Changes in the local gross domestic products (GDP) are a good indicator for the variation in the level of activity.

Structure of local economy refers to the weight that different economic sectors (or subsectors) represent to the overall local economy. Describing the structure of economy of a specific municipality would consist on identifying the sectors responsible for a larger share of gross value added or for employing more workers and which sectors are less relevant.

Economic situation of households translates the financial situation of individuals. This includes the employment situation, which has a significant impact in the local energy system given the difference in the daily habits between employed and unemployed citizens, and the individuals purchase power. On the latter, there are several studies that demonstrate the relation between private purchase power and energy use.

3.2.4. Climate Variability

Finally, the effects of differences in climate conditions have also been included in the natural evolution of the local energy system. These vary from year to year, and temperature differences lead to changes in energy needs (mostly for heating and cooling), which must be considered. It is assumed that these changes can be assessed through the variations in annual heating and cooling degree days.

3.3. Higher-Level Governance Framework

Higher-level governance framework refers to the characteristics of international, national, and regional governance frameworks existing within the studied timeframe and focusing on energy and climate issues. Considering the variables identified in the review, it is assumed to be appropriate to separate technology evolution from specific energy and climate policies.

The full list of factors that characterize higher-level governance framework is presented in Table 6, along with the respective characterizing indicators.

Table 6. Factors of change related to higher-level governance framework with corresponding indicators.

Factor of Change		Characterizing (and Quantifiable) Indicators	Unit
Technology Evolution	Technical evolution	National average energy intensity per sector	kWh/€
	Market evolution: Technology costs Market barriers	Over cost of A^{++} appliances compared to B equivalent Energy efficiency gap	€
	Energy supply sector: Fuel-mix of electricity and heat generation Efficiency of energy transformation sector	Share of renewable energy sources (RES) Overall efficiency and losses	Dimensionless (%) Dimensionless (%)
National and international energy-related policies	Existing policies	Public and private investments on environmental management and protection per economic activity (cumulative values)	€
	Energy costs	Average energy cost for final consumer	€/kWh
	Coherence with local policies	Alignment between different scales objectives?	(Y/N)
	Support of local action	Initiatives to promote local action (Number)	# initiatives

3.3.1. Technology Evolution

Technology evolution can be prompted by energy related policies (targeting energy efficiency and others) but it may also be market driven (following industry needs or an attempt to reduce costs). The evolution of the energy supply sector can be seen as a mix between natural technology evolution and an effect of energy and climate policies. The energy efficiency improvement on conversion plants is an example of the former, while the preference for renewable energy sources to produce electricity, due to fee-in tariffs in the technological maturation phase, is an example of the latter.

Technical evolution refers to the technology development that occurs naturally, i.e., excluding development that is promoted by energy efficiency policies and other related instruments. This evolution can be assessed by the variation in the energy intensity for each activity sector, at national or even international level.

Market evolution corresponds to the changes in the market that lead to changes in the technology choices by both individuals and companies. This can be translated into changes in technology costs, as well as the presence of market barriers that hamper the shift towards more efficient solutions. The variation in technology costs can be characterized by, e.g., the over cost of A++ appliances compared to a B equivalent. The presence of market barriers and respective evolution over time can be assessed by the quantification of the energy efficiency gap in different moments in time.

Energy supply sector represents the changes in the national and regional energy transformation and transport activities. It includes the evolution in electricity and heat production and transport, as well as the changes in the refinery of fossil fuels. Thus, it comprises the changes in the fuel-mix of electricity and heat generation, i.e., the increase/decrease in the share of renewable energy sources that are used in electricity production, as well as the changes in the efficiency of the energy supply sector (electricity and heat production, but also oil products transformation).

3.3.2. National and International Energy-Related Policies

On what concerns the factors of change associated to energy and climate policies, two distinct groups of factors were found: the ones that characterize the existing energy and climate related policies, e.g., in terms of implementation and specific targets, and the ones that characterize the existing institutional framework regarding the promotion and support of local actions towards climate change mitigation. It is also important to mention the influence that the local context may have on the overall effect of these factors of change on the local energy system. The effects of national and international policies depend on the reaction of targeted actors to the policy instruments put in place. In turn, their reaction depends on their level of education and awareness as well as cultural habits and financial situation. All the identified factors associated with energy and climate policies are briefly described below.

Existing policies correspond to the policies that are implemented at regional, national, and international level that have a climate and/or energy purpose. These may impact positively or negatively local GHG emissions, depending on the coherence between local policies and the ones defined at higher levels of governance. Here, it is assumed that the public and private investments on environmental management and protection could be a good indicator of the scale of existing policies at governance levels other than local.

Energy costs refer to the changes in the average cost of the different energy products and the impact that these may have in the local energy system. Energy prices are usually established at the national and/or regional level, and thus are included within the changes associated with the higher-level governance framework. The average unitary price per energy product in the consumer sector, as well as the respective evolution over time, is considered a good indicator to characterize this factor of change.

Coherence with local policies refers to whether national and regional policy objectives are aligned with those that were defined at the local/municipal level. Here, the quantification of this alignment is not feasible, thus a yes or no question is suggested.

Support of local action refers to the existence of initiatives promoted at national and international level that promote actions taken at the local level. This may include technical support for the development of action plans, and energy dedicated actions, and economic incentives and support (as low-rate loans). A possible indicator to characterize the level of support could be the number of initiatives promoted at higher levels of governance that support local action.

3.4. Local Climate Change Mitigation Actions

Finally, local climate change mitigation actions include any policy actions that are promoted at the local level, by (or with the support of) local authorities, with the aim of reducing local GHG emissions. These include actions taken under a formal and wider plan and also isolated actions with a very specific target (e.g., the improvement of energy efficiency in public lighting). Here, the factors of change that have been identified are mostly related with the characterization of these actions, in terms of objectives and targeted actors as well as in what concerns implementation procedures. The factors that were identified as important for the characterization of local climate change mitigation actions are briefly described below and presented in Table 7.

Table 7. Factors of change associated with local climate change mitigation actions with corresponding indicators.

Factor of Change		Characterizing (and Quantifiable) Indicators	Unit
Actors involvement	Action plan preparation	Groups of actors involved	Nominal variable
	Action plan implementation	Groups of actors involved	Nominal variable
Timeframe	Study	Base Year and Time Horizon	Year
	Action plan implementation	Beginning and End of action plan implementation	Year
Targets	Vision	Other goals besides GHG emissions reduction	(Y/N)
	GHG emissions reduction		ton $CO_{2\ eq.}$
Current implementation status	Overall	Share of the overall plan already implemented	Dimensionless (%)
	Per action	Share of the individual measure already implemented	Dimensionless (%)
Estimated impact	Base year inventory	In terms of final energy and GHG emissions reduction (measured)	GWh/a ton $CO_{2\ eq.}$/a
	BAU projections	In terms of final energy and GHG emissions reduction (predicted)	GWh/a ton $CO_{2\ eq.}$/a
	Latest inventory	In terms of final energy and GHG emissions reduction (measured)	GWh/a ton $CO_{2\ eq.}$/a
Budget	Overall		€
	Spent so far		€
Financing sources	Entity(s) responsible	Entities financing local actions (per category)	Nominal variable
	Share of overall investment costs	Fraction of overall investment that is financed by each entity	Dimensionless (%)
Actions characterization	Area of intervention	Sector and subsector targeted	
	Technical measure:		
	Measure	Building retrofit, modal shift in transports, led street lights, etc.	Nominal variable
	Level of implementation	Share of the sector/subsector impacted by the action	Dimensionless (%)
	Estimated impact	In terms of final energy and GHG emissions reduction	GWh/a ton $CO_{2\ eq.}$/a
	Policy instrument:		
	Type	Regulation, incentive, education, etc.	Nominal Variable
	Level of support	Share of financing provided by the instrument	Dimensionless (%)
	Stakeholders involvement		Nominal variable

Actors' involvement refers to the group of actors that were/are involved in the local climate change mitigation actions. A distinction is made between the actors that are involved in the planning process (where actions are defined) and the ones involved in the implementation stage. The involvement of different stakeholders in the planning and implementation of climate and energy actions is considered to be beneficial for the respective acceptance and adherence by local citizens. The characterization of this involvement may be achieved by the identification of the groups of actors that were involved (e.g., local authority, local/regional energy agency, external consultant, local NGOs, citizens, etc.) and their level of involvement (co-creation, draft version public consultation, etc.).

Timeframe refers to the time span of the actions being planned/implemented. Throughout the review, the consideration of the time span of the ex-ante planning exercise (when existing) was also noted as relevant for the assessment of the outcomes of local actions. This is translated into the beginning and end years of the actual (or projected) action implementation.

Targets refer to the objectives that were defined at the local level. These include the level of GHG emissions reductions that is aimed, the wider vision concerning the local energy system, and whether there are other goals besides GHG emissions (such as creation of jobs or decrease in energy costs).

Current implementation status refers to the degree of implementation of the wider action plan (if existing) and of the different individual actions that are projected. It can be assessed as a percentage of the overall implementation target. For instance, considering an action that implied the installation of solar thermal systems for water heating in 100 residential buildings, the degree of implementation would correspond to the identification of the share of systems that were already installed.

Estimated impact corresponds to the projected outcome of the proposed actions, in terms of both energy use and GHG emissions. When a mid-term assessment is also performed, the quantification of the already achieved reductions could also be performed.

Budget refers to the level of investment needed to implement the proposed actions. Here, it is important to distinguish between the overall projected budget and the investments made thus far.

Financing sources correspond to the entities responsible for financing the actions' implementation. This may comprise different entities, both public and private. The characterization of the financing sources could be performed by the identification of the type of entities financing the local actions (local authority's own resources, national funds and programs, EU funds and programs, other), and the level of investment of each of the different entities.

Actions characterization corresponds to the characterization of the individual actions planned and/or implemented at the local level, in terms of: (1) area of intervention; (2) type of technical measure; and (3) policy instrument. Area of intervention is the sector and subsector of the local energy system which will be impacted by the action (residential, services, transports, etc.). Then, the technical measure refers to the type of change that is aimed at, as well as the degree of change that is intended. For instance, the shift of 10% of private cars passenger-kilometers to soft modes of transportation could be seen as a technical measure. Finally, the choice of policy instrument corresponds to the type of mechanism that is used to achieve the desired changes in the system. Using the same example, this could comprise the creation of dedicated bike lanes and/or the implementation of a congestion charge. When assessing the policy instruments, it is also useful to identify the level of financial support that is provided to the adopters of the measures (i.e., if there are economic incentives to change) and which actors are involved in their implementation.

Similar to what was said regarding the effects of higher-level policies, the influence of local context on the contribution of local actions cannot be disregarded. Local inhabitants' characteristics may influence the outcomes of policy actions in a significant manner. Moreover, local authorities' characteristics are an important feature on the definition and implementation of local actions. Thus, even if local actions are similar in their characteristics, their outcome can be very different if the context in which they are implemented is not the same.

4. Discussion

There are several studies that analyze the changes in local GHG emissions and potential explanatory factors. However, it was noted throughout the review that existing work is usually constrained to the consideration of a specific set of factors of change, disregarding others that may also have a significant impact on local emissions. For instance, it is common to narrow the focus to the actions that are being taken at the local level, disregarding the influence of higher-level governance framework in which they are being implemented and that may have confounding effects.

Such non-consideration of other potential factors of change hinders the accuracy of the results, as the effects calculated and associated with the considered factors may turn out to be either under- or overestimated. Indeed, the evaluation of the effects of local climate change mitigation actions will not lead to accurate results unless the remaining factors of change are taken into account. For example, in [5], the impact of local climate actions was inconclusive, partly due to the non-consideration of factors associated with the changes in local socio-economic and cultural conditions. Moreover, a better understanding of all potential factors of change of local GHG emissions can lead to better planning at local as well as regional and national levels. The interaction between the policies implemented at different governance levels could be improved by a comprehensive consideration of all factors of change and their interactions.

It is worth noting the overly simplified consideration of the local context in most of the existing work. Some of the reviewed documents refer to its importance, and how the local context specificities can be determinant for the success of local actions implementation, yet quantification methods tend to be overlooked. The systematic identification of the factors of change associated with local context may thus be a step forward for their consideration in future empirical ex-post evaluation studies and/or ex-ante planning exercises.

5. Conclusions

This paper is dedicated to a comprehensive and systematic identification and characterization of the factors that are relevant for the assessment of the observed changes in GHG emissions associated with local energy systems. This is an important contribution for the assessment of the actual effects of local climate change mitigation actions, by establishing the bases for developing analytical frameworks.

The identification was performed through a systematic literature review process. The complete list of variables encountered in the review was reduced using eligibility criteria, as well as three screening filters (explicit repetition, implicit repetition, and dependent variables). A list of 107 variables that were grouped into 40 factors of change was finally obtained.

The factors of change were grouped into four subsets, corresponding to three sources of change plus the local context. The sources of changes subsets are: local socio-economic and cultural changes; higher-level governance framework; and local climate change mitigation actions. The local context corresponds to a group of variables that, despite not leading directly to changes in local GHG emissions, have influence on how the remaining factors impact the local energy system—e.g., local inhabitants' level of awareness and commitment towards environmental problems or local authorities' legal scope of action.

Even if the identification of the major categories has been previously accomplished in academic studies, the major advancement of this work corresponds to their disaggregation into individual factors of change and respective characterization.

The factors identified correspond to all the factors that may potentially lead to changes in local GHG emissions. The relevance of the effects associated with each factor of change will depend on the local system in question, as well as on the timeframe addressed. Moreover, even acknowledging the difficulty of empirical validation (and quantification) of each factor individually, the obtained results comprise an advancement from what was possible thus far, providing a basis for more accurate models and analyses of the evolution of the energy systems at the local level.

These results can be used to build an analytical framework to characterize and model the evolution of a local energy system, assessing the individual effect of each of the different explanatory factors. This will allow for more accurate energy planning at the local level, taking into account the potential interactions with the external factors of change (i.e., the factors that are not controlled by local authorities). For instance, it may be used for ex-post evaluation of local climate change mitigation actions, allowing for the disaggregation of the observed changes in local GHG emissions into the causes and the isolation of the effects that were solely associated to local policies and actions.

The outcome of this work can be seen as a fundamental step towards a comprehensive assessment of the evolution of local energy systems. Such assessment is crucial for a better evaluation of the actual contribution of local climate-related policies and actions towards climate change mitigation.

Author Contributions: Conceptualization, I.A. and V.L.; Formal analysis, I.A. and V.L.; Funding acquisition, I.A.; Investigation, I.A.; Methodology, I.A.; Supervision, V.L.; Validation, I.A. and V.L.; Visualization, I.A.; Writing—original draft, I.A.; and Writing—review and editing, I.A. and V.L. All authors have read and agreed to the published version of the manuscript.

Funding: This research was funded by Fundação para a Ciência e Tecnologia (FCT—Portugal), grant number PD/BD/105862/2014, in the frame of the MIT Portugal Program. INEGI (LAETA—Grupo de Energia e Ambiente) provided financial support with the article processing charges.

Conflicts of Interest: The authors declare no conflict of interest.

Appendix A

This appendix presents in detail the process of identification and screening of the factors of change that are relevant for local GHG emissions. Table A1 presents the full list of variables identified in the literature. The four following tables are dedicated to the screening process: Table A2 presents the variables that were excluded for non-applicability; Tables A3 and A4 present the variables that were eliminated due to explicit and implicit repetition, respectively; and Table A5 presents the variables that were excluded to avoid dependent variables.

Table A1. Full list of variables identified in the systematic literature review process and respective sources.

Document	Code	Variable
Auld et al. (2014)	1.001	Agenda setting processes
	1.002	Favorable public opinion
	1.003	Favorable party platform
	1.004	Recent change in government
	1.005	Presence of champions
	1.006	Social appeal
	1.007	Availability of technology
	1.008	International processes
	1.009	Focusing events
	1.010	Action of feedback mechanism
	1.011	Timeframe
	1.012	Reporting nature
	1.013	Policy instruments
	1.014	State of activity
	1.015	Target of policy (actors)
	1.016	Source of authority at different stages (Hybrid or Government)
	1.017	Presence of flexibility mechanisms
Boonekamp (2004)	2.001	GDP growth
	2.002	Structure, main sectors
	2.003	Structure, subsectors
	2.004	Structure, final demand
	2.005	Savings, final demand
	2.006	Conversion efficiency end-users
	2.007	Co-generation end-users
	2.008	Fuel mix end-users
	2.009	Export
	2.010	Structure electricity production sectors
	2.011	Co-generation electricity production sectors
	2.012	Import
	2.013	Fuel mix electricity production sectors

Table A1. *Cont.*

Document	Code	Variable
Boonekamp (2005)	3.001	Energy statistics (statistical figures)
	3.002	Corrected statistics (corrections from trend breaks, etc.)
	3.003	Climate (degree-days and fraction room heating)
	3.004	GDP growth (GDP growth rate)
	3.005	Structure, main sectors (growth rate main sectors)
	3.006	Structure, subsectors (growth rate per subsector)
	3.007	Structure, final demand (Reference final demand per subsector)
	3.008	Savings, final demand (Actual final heat/electricity demand)
	3.009	Conversion efficiency end-users (Actual efficiency boilers, etc.)
	3.010	Cogeneration end-users (actual CHP figures)
	3.011	Fuel-mix end-users (Actual energy use)
	3.012	Export (Actual energy exports)
	3.013	Structure energy transformation sectors (reference use refineries/gas supply)
	3.014	Cogeneration energy transformation sectors (actual cogeneration figures)
	3.015	Efficiency energy transformation sectors (actual conversion efficiencies)
	3.016	Import (actual energy imports)
	3.017	Fuel-mix electricity production sectors (actual energy use)
Broto and Bulkeley (2013)	4.001	Country
	4.002	Urban area
	4.003	Population
	4.004	Density
	4.005	GDP
	4.006	GDP per capita
	4.007	World City Rank
	4.008	Annual population rank
	4.009	Location
	4.010	Dates
	4.011	Urban character
	4.012	Type of experiment
	4.013	Objectives
	4.014	Type of innovation
	4.015	Institutional factors
	4.016	Sector specific information
	4.017	Actors involved
	4.018	Funding
	4.019	Mode of governance
	4.020	Environmental justice
	4.021	Action targeted sectors
	4.022	Action type of schemes
CoM (2010)	5.001	Overall CO_2 reduction target
	5.002	Vision
	5.003	Coordination and organizational structures created/assigned
	5.004	Staff capacity allocated
	5.005	Involvement of stakeholders and citizens
	5.006	Overall estimated budget for the Sustainable Energy Action Plan implementation
	5.007	Foreseen financing sources for the implementation
	5.008	Monitoring process
	5.009	Year
	5.010	Population
	5.011	Final energy consumption per sector and carrier
	5.012	Energy supply sources
	5.013	CO_2 emissions per sector and carrier
	5.014	Date of approval
	5.015	Decision body approving the plan
	5.016	Business-as-usual projections

Table A1. *Cont.*

Document	Code	Variable
CoM (2010)	5.017	Area of intervention
	5.018	Policy instrument
	5.019	Origin of the action (local or other)
	5.020	Responsible body
	5.021	Implementation timeframe
	5.022	Estimated implementation cost
	5.023	Energy savings
	5.024	Renewable energy production
	5.025	CO_2 reduction
	5.026	Newest inventory year
	5.027	Population
	5.028	Final energy consumption per sector and carrier
	5.029	Energy supply sources
	5.030	CO_2 emissions per sector and carrier
	5.031	Status implementation
	5.032	Implementation cost spent so far
	5.033	Renewable energy production from actions implemented so far
	5.034	CO_2 reduction from actions implemented so far
Corfee-Morlot et al. (2009)	6.001	Coherent multi-level policy framework (vertically and horizontally)
	6.002	Existence of a dedicated institution/agency to deal with climate change mitigation actions/plans
	6.003	Local authority's capacity and expertise on climate change mitigation
	6.004	Local authority's ability to obtain funding
	6.005	Local authority's responsibility and competences
	6.006	Support from central governments to local policies
Farla and Blok (2000)	7.001	Total dwelling area
	7.002	Average number of inhabitants
	7.003	Number of appliances per type
	7.004	Number of vehicle-kilometers of personal cars per fuel
	7.005	Number of passenger-kilometers per mode (except car)
	7.006	Number of vehicle-kilometers of freight road transportation
	7.007	Number of ton-kilometers per mode (other freight transportation)
	7.008	Number of employee-years
	7.009	Production of goods per type (piece, kg or m^2)
	7.010	Cultivated land area
	7.011	Raising of cattle, pigs and chickens
	7.012	Volume index of production
Fortes et al. (2014)	8.001	Gross domestic product (GDP)
	8.002	Population
	8.003	Private consumption
	8.004	Gross value-added Agriculture
	8.005	Gross value-added Services
	8.006	Gross value-added Transports
	8.007	Gross value-added Industry
Fortes et al. (2015)	9.001	Annual growth rate of the socio-economic indicator
	9.002	Demand elasticity
	9.003	Autonomous efficiency improvement factor (industrial sectors)
	9.004	Energy and environmental policies
	9.005	Energy and environmental policy instruments
	9.006	Technical and costs evolution
	9.007	Availability and capacity limits
	9.008	Other information (discount rate, etc.)
	9.009	Endogenous resources potential and prices
	9.010	Import/export prices and boundaries

Table A1. *Cont.*

Document	Code	Variable
Gustavsson et al. (2009)	10.001	Actors involved in local Climate Change mitigation
	10.002	Formulation of the climate-policy strategy
	10.003	Results in terms of GHG emissions reduction
	10.004	Geography
	10.005	Business structure
	10.006	Physical circumstances
	10.007	Actors involvement
Gysen et al. (2002)	11.001	Objectives
	11.002	Instruments
	11.003	Means—Societal
	11.004	Means—Policy process
	11.005	Policy output
	11.006	Policy outcome
	11.007	Policy impact
IAEA (2006)	12.001	Total population
	12.002	Average annual growth rate of population
	12.003	Share of urban population
	12.004	Average household size in urban areas
	12.005	Average household size in rural areas
	12.006	Share of population of age 15–64 in total population
	12.007	Share of potential labor force actually working
	12.008	Share of population living in large cities
	12.009	GDP
	12.010	Average annual growth rate of GDP
	12.011	Distribution of GDP formation by kind of economic activity
	12.012	Distribution of sector gross value-added by subsectors
	12.013	Energy intensities in industry
	12.014	Penetration of energy carriers into useful thermal energy demand of the different sectors
	12.015	Average efficiencies of fuels for thermal uses per sector
	12.016	Share of transportation modes in the total demand for freight transportation
	12.017	Energy intensity of freight transportation modes
	12.018	Average intracity distance traveled per person per day
	12.019	Average load factor of intracity passenger transportation mode(s)
	12.020	Share of transportation mode(s) in the total demand for intracity passenger transportation
	12.021	Energy intensity of transportation mode(s) in intracity travel
	12.022	Average intercity distance traveled per person per day
	12.023	Inverse of car ownership ratio
	12.024	Average intercity distance driven per car per year
	12.025	Average load factor of intercity transportation mode(s)
	12.026	Share of car type(s) in the intercity transportation mode(s)
	12.027	Share of public transportation mode(s) in the intercity passenger travel by public modes
	12.028	Energy intensity of transportation mode(s)
	12.029	Fraction of urban dwellings in areas where space heating is required
	12.030	Degree days for urban dwellings
	12.031	Fraction of urban dwellings per type
	12.032	Average size of urban dwellings by type
	12.033	Fraction of floor area that is actually heated in urban areas per dwelling type
	12.034	Specific heat loss rate by urban dwelling type
	12.035	Share of urban dwellings with air conditioning, by dwelling type
	12.036	Specific cooling requirements by urban dwelling type
	12.037	Specific energy consumption for cooking in urban dwellings
	12.038	Share of urban dwellings with hot water facilities
	12.039	Specific energy consumption per urban dwelling for electric appliances

Table A1. *Cont.*

Document	Code	Variable
IAEA (2006)	12.040	Electricity penetration for appliances in urban households
	12.041	Specific fossil fuel consumption per urban dwelling for lighting and non-electric appliances
	12.042	Penetration of various energy forms into space heating in urban households
	12.043	Efficiency of various fuels use, relative to that of electricity use, for space heating in urban households
	12.044	Coefficient of performance of heat pumps for space heating in urban households
	12.045	Penetration of various energy forms into water heating in urban households
	12.046	Efficiency of various fuels use, relative to that of electricity use, for water heating in urban households
	12.047	Coefficient of performance of heat pumps for water heating in urban households
	12.048	Approximate share of water heating demand in urban households that can be met with solar installations
	12.049	Penetration of various energy forms into cooking in urban households
	12.050	Efficiency of various fuels use, relative to that of electricity, for cooking in urban households
	12.051	Approximate share of cooking demand in urban households that can be met with solar installations
	12.052	Share of air conditioning demand of urban households that can be met with electricity
	12.053	Coefficient of performance of electric air conditioning in urban households
	12.054	Coefficient of performance of non-electric air conditioning in urban households
	12.055	Fraction of rural dwellings in areas where space heating is required
	12.056	Degree days for rural dwellings
	12.057	Fraction of rural dwellings per type
	12.058	Average size of rural dwellings by type
	12.059	Fraction of floor area that is actually heated in rural areas per dwelling type
	12.060	Specific heat loss rate by rural dwelling type
	12.061	Share of rural dwellings with air conditioning, by dwelling type
	12.062	Specific cooling requirements by rural dwelling type
	12.063	Specific energy consumption for cooking in rural dwellings
	12.064	Share of rural dwellings with hot water facilities
	12.065	Specific energy consumption per rural dwelling for electric appliances
	12.066	Electricity penetration for appliances in rural households
	12.067	Specific fossil fuel consumption per rural dwelling for lighting and non-electric appliances
	12.068	Penetration of various energy forms into space heating in rural households
	12.069	Efficiency of various fuels use, relative to that of electricity use, for space heating in rural households
	12.070	Coefficient of performance of heat pumps for space heating in rural households
	12.071	Penetration of various energy forms into water heating in rural households
	12.072	Efficiency of various fuels use, relative to that of electricity use, for water heating in rural households
	12.073	Coefficient of performance of heat pumps for water heating in rural households
	12.074	Approximate share of water heating demand in rural households that can be met with solar installations
IAEA (2006)	12.075	Penetration of various energy forms into cooking in rural households
	12.076	Efficiency of various fuels use, relative to that of electricity, for cooking in rural households
	12.077	Approximate share of cooking demand in rural households that can be met with solar installations

Table A1. *Cont.*

Document	Code	Variable
IAEA (2006)	12.078	Share of air conditioning demand of rural households that can be met with electricity
	12.079	Coefficient of performance of electric air conditioning in rural households
	12.080	Coefficient of performance of non-electric air conditioning in rural households
	12.081	Share of service sector in the total active labor force
	12.082	Average floor area per employee in the Service sector
	12.083	Share of Service sector floor area requiring space heating
	12.084	Share of Service sector floor area requiring space heating that is actually heated
	12.085	Specific space heat requirements of Service sector floor area
	12.086	Share of air-conditioned Service sector floor area
	12.087	Specific cooling requirements in the Service sector
	12.088	Energy intensity of motor fuel use per Services' subsector
	12.089	Energy intensity of electricity specific uses per Services' subsector
	12.090	Energy intensity of thermal uses (except space heating) per Services' subsector
	12.091	Penetration of various energy forms into space heating in Service sector
	12.092	Penetration of various energy forms into other thermal uses in the Service sector
	12.093	Efficiency of various fuels use, relative to that of electricity use, for thermal uses in Service sector
	12.094	Coefficient of performance of heat pumps in space heating in Service sector
	12.095	Share of low-rise buildings in the total Service sector floor area
	12.096	Approximate share of thermal uses in the Service sector that can be met by solar installations
	12.097	Share of air-conditioning that can be met with electricity
	12.098	Coefficient of performance of electric air conditioning in Service sector
	12.099	Coefficient of performance of non-electric air conditioning in Service sector
IEA (2014)	13.001	Population
	13.002	Floor area per capita
	13.003	Persons per household
	13.004	Appliances ownership per capita
	13.005	Heat per floor area
	13.006	Domestic how water energy per capita
	13.007	Cooking energy per capita
	13.008	Electricity for lighting per floor area
	13.009	Energy per appliance
	13.010	Passenger-kilometers
	13.011	Share of total passenger-kilometers by mode
	13.012	Energy per passenger-kilometer by mode
	13.013	Tons-kilometers
	13.014	Share of total ton-kilometers by mode
	13.015	Energy per ton-kilometer by mode
	13.016	Services value-added
	13.017	Energy per value-added (services)
	13.018	Industry subsectors value-added
	13.019	Share of total value-added by subsector
	13.020	Energy per value-added by subsector
	13.021	Agriculture value-Added
	13.022	Share of total value-added by subsector
	13.023	Energy per value-added by subsector
	13.024	Changes in CO_2 intensity
	13.025	Changes in input coefficients
	13.026	Changes in the composition of final demand
	13.027	Changes in the level of final demand (economic growth)

Table A1. *Cont.*

Document	Code	Variable
Jovanovic et al. (2010)	14.001	Population growth
	14.002	Persons per household
	14.003	Share of the potential labor force
	14.004	Share of the population outside the community
	14.005	Share of rural population
	14.006	GDP
	14.007	GDP growth rate
	14.008	Monetary values per capita of the major sectors and subsectors
	14.009	Efficiency of appliances/vehicles
	14.010	Buildings insulation
Kasperson et al. (1995)	15.001	Population growth
	15.002	Migration
	15.003	Natural growth
	15.004	Irrigation development
	15.005	Mechanization
	15.006	Fertilization
	15.007	Pasture improvement
	15.008	Industrialization infrastructure
	15.009	Agricultural intensification
	15.010	Urbanization
	15.011	Poverty
	15.012	International market
	15.013	National market
	15.014	State policy
	15.015	Shift to commodity production
	15.016	Foreign debt, balance of trade
	15.017	Resource allocation rules and institutions
	15.018	Capital extraction
	15.019	Political corruption
	15.020	Frontier development
	15.021	Ethnic/religious views
	15.022	Mass-consuming view of nature
	15.023	Acceptance of corruption
Lin et al. (2010)	16.001	Population
	16.002	Population growth rate
	16.003	Size of households
	16.004	Number of households
	16.005	GDP
	16.006	Growth rate of GDP
	16.007	Measures and policies already promulgated
	16.008	Measure/Action
	16.009	Intensity of measure (degree of implementation)
	16.010	Competences of local authorities in implementation fields
	16.011	Investment costs
	16.012	Implementation experiences
Millard-Ball (2012)	17.001	Population
	17.002	Employment
	17.003	Average income
	17.004	Education
	17.005	Environmental voting
	17.006	Civic
	17.007	Non-residential building
	17.008	Latitude
	17.009	Ped/bike mode share
	17.010	Time

Table A1. *Cont.*

Document	Code	Variable
Millard-Ball (2012)	17.011	Number of certified buildings
	17.012	Share of renewable energy sources electricity microgeneration
	17.013	Street lighting expenditure
	17.014	Use of soft modes
Morlet and Keirstead (2013)	18.001	City power (EU, 2007)
	18.002	City size (population)
	18.003	Legal structure and status
	18.004	Spending power
	18.005	Control over income
	18.006	Civic involvement (voting patterns)
	18.007	Herfindahl–Hirschman index (degree of competition in the national electricity system)
	18.008	Degree of electricity generated in combined heat and power systems
	18.009	City's population
	18.010	Wealth (GDP per capita)
	18.011	Climate (Heating Degree Days)
	18.012	Price of energy (gas and electricity) and relative share of any taxes or levies—energy costs
	18.013	Carbon intensity of fuels
	18.014	Targets set by each city for GHG emissions reductions
	18.015	Annual energy demand
	18.016	Policy type (Financial/Non-financial)
Murphy et al. (2012)	19.001	Policy instrument combinations
	19.002	Obligating/Incentivizing balance
	19.003	Long-term program
	19.004	Non-generic instruments
	19.005	Primacy to energy efficiency
	19.006	Whole house/deep retrofit
	19.007	Energy sufficiency
	19.008	Policy theory associated with instruments applied
	19.009	Description of the impact of instruments (secondary sources and interviews)
Nassen (2014)	20.001	Total consumption
	20.002	Household size
	20.003	Age of occupants
	20.004	Education of occupants
	20.005	Dwelling type
	20.006	Urbanity
Neji and Astrand (2006)	21.001	Changes in knowledge
	21.002	Awareness and behavior of important actors
	21.003	Technology performance
	21.004	Price development
	21.005	Sales data
	21.006	Market share
	21.007	Changes in manufacturers assortment
Praznik et al. (2013)	22.001	Surface area to volume ratio
	22.002	Orientation of the facades' transparent elements
	22.003	Thermal zoning of the building and the separation of its unheated parts from the thermal envelope
	22.004	Heated area per occupant
Rai and Robinson (2015)	23.001	Attitude factors
	23.002	Economic factors
	23.003	Social factors (impact on agents' attitude)

Table A1. *Cont.*

Document	Code	Variable
Schipper et al. (2001)	24.001	Population
	24.002	Floor area per capita
	24.003	Persons per household
	24.004	Appliances ownership per capita
	24.005	Intensity for different uses
	24.006	Total passenger-kilometers
	24.007	Distribution of passenger-kilometers per mode
	24.008	Energy/passenger-kilometer per mode
	24.009	Distribution of value-added per subsector
	24.010	Average energy/value-added per subsector
Sobrino and Monzon (2014)	25.001	Carbon intensity of road transport
	25.002	Energy intensity of road transport
	25.003	Use intensity
	25.004	Motorization rate
	25.005	Job intensity
	25.006	Workers income intensity
	25.007	GDP
Spyridaki and Flamos (2014)	26.001	Electricity market structure and design
	26.002	Interconnection of domestic electricity markets in Europe
	26.003	Primary factor: labor, capital, conventional and non-conventional resources
	26.004	Supply and demand parameters
	26.005	Incumbent policy framework included in the reference scenario
	26.006	Different renewable energy sources penetration levels
	26.007	Renewable energy source potential
	26.008	Efficiency of the whole electricity system
	26.009	Emissions of other air pollutants
	26.010	Discount rate for investments
	26.011	Firm behavior
	26.012	Allowance trade patterns among regions
	26.013	Socio-demographic and lifestyle trends
	26.014	Interconnection of domestic electricity markets in Europe
	26.015	Political context in which policy instruments are imposed
	26.016	Supply and demand parameters
	26.017	Firm behavior
	26.018	Simplification in technology production options and load segments of electricity production
	26.019	Limited analysis of constraints in output
	26.020	Slope of demand and supply curves
	26.021	Electricity market structure and design
	26.022	Technology market failures and other externalities related to electricity generation design
	26.023	Trading options
	26.024	Banking of emissions
	26.025	Alternative compliance payments
	26.026	Distinctions between price based or quantity-based policy instruments with variations in prices and quotas for commodities
	26.027	Uniform/Differentiated feed in tariff rate
	26.028	Phase in of Renewable Portfolio Standard
	26.029	Annual digression rate of feed in tariff rate
	26.030	Limitation of the payment period
	26.031	Tariff reduction due to inflation
	26.032	Distinctions between price based or quantity-based policy instruments with variations in prices and quotas for commodities
	26.033	International emissions trading
	26.034	Uniform and unilateral imposition of carbon taxes across all EU-ETS regions
	26.035	Lump-sum treatment of additional tax revenues

Table A1. *Cont.*

Document	Code	Variable
Spyridaki and Flamos (2014)	26.036	Stringency levels
	26.037	Different application scope
	26.038	Nature of targets, the target groups, the policy-implementing agents, the available budget, the available information on the initially expected energy-savings impact, and the cost effectiveness of the instrument
Spyridaki and Flamos (2014)	26.039	Distinctions between price-based or quantity-based policy instruments
	26.040	Different renewable energy sources and support design elements
	26.041	Variable scenarios in the short and long run for key policy parameters such as price of certificate, level of obligation, level of sales tax and the level of penalty
	26.042	Fixed-price policies and endogenous price policies
	26.043	Technology specific hurdle rates reflecting market barriers, consumer preferences and risk factors limiting purchase of new energy technologies
	26.044	Conditions for implementation and proper utilization of saving options (technology equipment availability, familiarity with the policy, overcoming barriers, motivation to invest)
	26.045	Specific implementation of policy instruments with regard to their funding
	26.046	Transaction costs
	26.047	Stability and credibility in policy regime
	26.048	Implementation period of the policy instrument
	26.049	Circumstances in which to apply a policy instrument (challenges in addressing different target groups, challenges in addressing different scopes, addressing a financial, knowledge barrier, internalizing externalities, addressing market competition, conditions under which a policy combination is required, conditions of policy redundancy)
	26.050	Implicit and explicit assumptions in the policy implementation process and mapping the cause-impact relationships
	26.051	Transaction costs related to the combined policy cycle
	26.052	Regulatory decisions on the additionality of energy savings from individual projects
Tang et al. (2010)	27.001	Political will
	27.002	State-level mandates concerning climate change (national)
	27.003	Wealth
	27.004	Coastal distance
	27.005	Population density
	27.006	Hazard damage
	27.007	Energy consumption
	27.008	Light transportation (soft modes and public transportation)
	27.009	Average commuting time
	27.010	Vehicular emission
	27.015	Concept of climate change or global warming
	27.016	Concept of GHG emission
	27.017	Effects and impact of climate change
	27.018	Long-term goals and detailed targets for GHG emissions
	27.019	Emissions inventory
	27.020	Base year emissions
	27.021	Emissions trend forecast
	27.022	Vulnerability assessment
	27.023	Cost estimates for GHG emissions reduction
	27.024	Using analysis tools
	27.025	Public awareness, education, and participation
Tang et al. (2010)	27.026	Inter-organizational coordination procedures
	27.027	GHG emissions reduction fee
	27.028	Establish a carbon tax
	27.029	Disaster resistant land use and building code
	27.030	Mixed use and compact development

Table A1. *Cont.*

Document	Code	Variable
	27.031	Infill development and reuse of remediated brownfield sites
	27.032	Green building and green infrastructure
	27.033	Low-impact design for impervious surface
	27.034	Control of urban service/growth boundaries
	27.035	Alternative transportation strategies
	27.036	Transit-oriented development and corridor improvements
	27.037	Parking standards adjustment
	27.038	Pedestrian/resident friendly, bicycle friendly, transit-oriented community design
	27.039	Renewable energy and solar energy
	27.040	Energy efficiency and energy stars
	27.041	Landfill methane capture strategies
	27.042	Zero waste reduction and high recycling strategy
	27.043	Creation of conservation zones or protected areas
	27.044	Watershed-based and ecosystem-based land management
	27.045	Vegetation protection
	27.046	Establish implementation priorities for actions
	27.047	Financial/budget commitment
Tang et al. (2010)	27.048	Identify roles and responsibilities among sectors and stakeholders
	27.049	Continuous monitoring, evaluation and update
	27.050	Overall energy efficiency gains
	27.051	Final intensity
	27.052	Primary intensity
	27.053	Energy efficiency gains
	27.054	Energy intensity (industry, manufacturing, chemicals, European Union structure)
	27.055	Specific consumption (steel, cement, paper)
	27.056	Energy efficiency gains
	27.057	Consumption per unit of traffic (road car equivalent, freight, air)
	27.058	Car efficiency (fleet average, new cars)
	27.059	Mobility in transport (public transport, freight)
	27.060	Energy efficiency gains
	27.061	Consumption per dwelling (total, electricity, scaled to European Union climate)
	27.062	Heating consumption (per dwelling, per m^2)
	27.063	Energy consumption per employee (total, electricity)
	27.064	Energy intensity (total, electricity)

Table A2. Variables excluded for non-applicability in the systematic literature review process.

Code	Variable
1.004	Recent change in government
1.005	Presence of champions
1.008	International processes
1.009	Focusing events
1.010	Action of feedback mechanism
4.007	World City Rank
4.009	Location
9.002	Demand elasticity
9.009	Endogenous resources potential and prices
9.010	Import/export prices and boundaries
10.003	Results in terms of GHG emissions reduction
10.004	Geography
10.005	Business structure
10.006	Physical circumstances

Table A2. *Cont.*

Code	Variable
11.005	Policy output
11.006	Policy outcome
11.007	Policy impact
12.008	Share of population living in large cities
12.037	Specific energy consumption for cooking in urban dwellings
12.039	Specific energy consumption per urban dwelling for electric appliances
12.041	Specific fossil fuel consumption per urban dwelling for lighting and non-electric appliances
12.063	Specific energy consumption for cooking in rural dwellings
12.065	Specific energy consumption per rural dwelling for electric appliances
12.067	Specific fossil fuel consumption per rural dwelling for lighting and non-electric appliances
15.006	Fertilization
15.007	Pasture improvement
15.015	Shift to commodity production
15.016	Foreign debt, balance of trade
15.017	Resource allocation rules and institutions
15.018	Capital extraction
15.019	Political corruption
15.020	Frontier development
15.023	Acceptance of corruption
17.007	Non-residential building
17.008	Latitude
17.011	Number of certified buildings
17.013	Street lighting expenditure
18.007	Herfindahl–Hirschman index (degree of competition in the national electricity system)
18.013	Carbon intensity of fuels
18.015	Annual energy demand
20.001	Total consumption
21.007	Changes in manufacturers assortment
22.001	Surface area to volume ratio
22.002	Orientation of the facades' transparent elements
22.003	Thermal zoning of the building and the separation of its unheated parts from the thermal envelope
22.004	Heated area per occupant
26.001	Electricity market structure and design
26.002	Interconnection of domestic electricity markets in Europe
26.003	Primary factor: labor, capital, conventional and non-conventional resources
26.004	Supply and demand parameters
26.005	Incumbent policy framework included in the reference scenario
26.007	Renewable energy source potential
26.009	Emissions of other air pollutants
26.011	Firm behavior
26.012	Allowance trade patterns among regions
26.014	Interconnection of domestic electricity markets in Europe
26.016	Supply and demand parameters
26.017	Firm behavior
26.019	Limited analysis of constraints in output
26.020	Slope of demand and supply curves
26.049	Circumstances in which to apply policy instrument (challenges in addressing different target groups, challenges in addressing different scopes, addressing a financial, knowledge barrier, internalizing externalities, addressing market competition, conditions under which a policy combination is required, conditions of policy redundancy)
26.050	Implicit and explicit assumptions in the policy implementation process and mapping the cause-impact relationships
26.051	Transaction costs related to the combined policy cycle
26.052	Regulatory decisions on the additionality of energy savings from individual projects
27.004	Coastal distance

Table A2. *Cont.*

Code	Variable
27.006	Hazard damage
27.007	Energy consumption
27.019	Emissions inventory
27.020	Base year emissions
27.022	Vulnerability assessment
27.024	Using analysis tools
27.062	Heating consumption (per dwelling, per m^2)
27.063	Energy consumption per employee (total, electricity)

Table A3. Variables excluded for explicit repetition in the systematic literature review process.

Code	Variable
1.011	Timeframe
1.013	Policy instruments
1.015	Target of policy (actors)
1.016	Source of authority at different stages (Hybrid or Government)
2.002	Structure, main sectors
2.003	Structure, subsectors
2.004	Structure, final demand
2.005	Savings, final demand
2.006	Conversion efficiency end-users
2.007	Co-generation end-users
2.008	Fuel mix end-users
2.009	Export
2.010	Structure electricity production sectors
2.011	Co-generation electricity production sectors
2.012	Import
2.013	Fuel mix electricity production sectors
3.004	GDP growth (GDP growth rate)
4.003	Population
4.004	Density
4.005	GDP
4.010	Dates
4.011	Urban character
4.017	Actors involved
4.020	Environmental justice
5.010	Population
5.027	Population
7.002	Average number of inhabitants
9.001	Annual growth rate of the socio-economic indicator
9.008	Other information (discount rate, etc.)
11.001	Objectives
11.002	Instruments
12.009	GDP
12.010	Average annual growth rate of GDP
12.011	Distribution of GDP formation by kind of economic activity
12.013	Energy intensities in industry
13.001	Population
13.002	Floor area per capita
13.003	Persons per household
13.004	Appliances ownership per capita
13.016	Services value-added
13.018	Industry subsectors value-added
13.021	Agriculture value-Added
14.006	GDP

Table A3. *Cont.*

Code	Variable
16.001	Population
16.002	Population growth rate
16.003	Size of households
16.005	GDP
16.006	Growth rate of GDP
16.011	Investment costs
17.001	Population
17.006	Civic
17.010	Time
17.014	Use of soft modes
18.002	City size (population)
18.009	City's population
18.010	Wealth (GDP per capita)
18.016	Policy type (Financial/Non-financial)
20.004	Education of occupants
20.006	Urbanity
24.001	Population
24.003	Persons per household
24.006	Total passenger-kilometers
24.007	Distribution of passenger-kilometers per mode
24.008	Energy/passenger-kilometer per mode
24.009	Distribution of value-added per subsector
25.007	GDP
26.008	Efficiency of the whole electricity system
26.048	Implementation period of the policy instrument
27.008	Light transportation (soft modes and public transportation)
27.023	Cost estimates for GHG emissions reduction
27.047	Financial/budget commitment
27.049	Continuous monitoring, evaluation and update
27.059	Mobility in transport (public transport, freight)
27.061	Consumption per dwelling (total, electricity, scaled to European Union climate)

Table A4. Variables excluded for implicit repetition in the systematic literature review process.

Code	Variable
1.003	Favorable party platform
1.007	Availability of technology
1.012	Reporting nature
3.005	Structure, main sectors (growth rate main sectors)
3.006	Structure, subsectors (growth rate per subsector)
4.001	Country
4.002	Urban area
4.006	GDP per capita
4.008	Annual population rank
4.013	Objectives
4.016	Sector specific information
4.022	Action type of schemes
5.005	Involvement of stakeholders and citizens
5.009	Year
7.001	Total dwelling area
7.003	Number of appliances per type
7.004	Number of vehicle-kilometers of personal cars per fuel
7.005	Number of passenger-kilometers per mode (except car)
7.006	Number of vehicle-kilometers of freight road transportation

Table A4. *Cont.*

Code	Variable
7.007	Number of ton-kilometers per mode (other freight transportation)
7.008	Number of employee-years
8.002	Population
9.003	Autonomous efficiency improvement factor (industrial sectors)
10.002	Formulation of the climate-policy strategy
10.007	Actors involvement
11.003	Means—Societal
12.001	Total population
12.002	Average annual growth rate of population
12.004	Average household size in urban areas
12.005	Average household size in rural areas
12.006	Share of population of age 15–64 in total population
12.016	Share of transportation modes in the total demand for freight transportation
12.017	Energy intensity of freight transportation modes
12.020	Share of transportation mode(s) in the total demand for intracity passenger transportation
12.021	Energy intensity of transportation mode(s) in intracity travel
12.027	Share of public transportation mode(s) in the intercity passenger travel by public modes
12.028	Energy intensity of transportation mode(s)
12.029	Fraction of urban dwellings in areas where space heating is required
12.030	Degree days for urban dwellings
12.031	Fraction of urban dwellings per type
12.032	Average size of urban dwellings by type
12.033	Fraction of floor area that is actually heated in urban areas per dwelling type
12.034	Specific heat loss rate by urban dwelling type
12.036	Specific cooling requirements by urban dwelling type
12.038	Share of urban dwellings with hot water facilities
12.055	Fraction of rural dwellings in areas where space heating is required
12.056	Degree days for rural dwellings
12.057	Fraction of rural dwellings per type
12.058	Average size of rural dwellings by type
12.059	Fraction of floor area that is actually heated in rural areas per dwelling type
12.060	Specific heat loss rate by rural dwelling type
12.062	Specific cooling requirements by rural dwelling type
12.064	Share of rural dwellings with hot water facilities
13.019	Share of total value-added by subsector
13.022	Share of total value-added by subsector
13.024	Changes in CO_2 intensity
13.025	Changes in input coefficients
13.026	Changes in the composition of final demand
13.027	Changes in the level of final demand (economic growth)
14.005	Share of rural population
14.007	GDP growth rate
14.008	Monetary values per capita of the major sectors and subsectors
14.009	Efficiency of appliances/vehicles
15.009	Agricultural intensification
15.010	Urbanization
16.004	Number of households
16.010	Competences of local authorities in implementation fields
17.009	Ped/bike mode share
17.012	Share of renewable energy sources electricity microgeneration
18.001	City power (EU, 2007)
18.008	Degree of electricity generated in combined heat and power systems
18.011	Climate (Heating Degree Days)
18.014	Targets set by each city for GHG emissions reductions

Table A4. *Cont.*

Code	Variable
19.001	Policy instrument combinations
19.002	Obligating/Incentivizing balance
19.003	Long-term program
19.004	Non-generic instruments
19.005	Primacy to energy efficiency
19.006	Whole house/deep retrofit
19.007	Energy sufficiency
20.002	Household size
21.001	Changes in knowledge
21.002	Awareness and behavior of important actors
21.003	Technology performance
21.006	Market share
25.001	Carbon intensity of road transport
25.002	Energy intensity of road transport
25.003	Use intensity
25.006	Workers income intensity
26.018	Simplification in technology production options and load segments of electricity production
26.023	Trading options
26.024	Banking of emissions
26.025	Alternative compliance payments
26.026	Distinctions between price based or quantity-based policy instruments with variations in prices and quotas for commodities
26.027	Uniform/Differentiated feed in tariff rate
26.028	Phase in of Renewable Portfolio Standard
26.029	Annual digression rate of feed in tariff rate
26.030	Limitation of the payment period
26.031	Tariff reduction due to inflation
26.032	Distinctions between price based or quantity-based policy instruments with variations in prices and quotas for commodities
26.033	International emissions trading
26.034	Uniform and unilateral imposition of carbon taxes across all EU-ETS regions
26.035	Lump-sum treatment of additional tax revenues
26.036	Stringency levels
26.037	Different application scope
26.038	Nature of targets, the target groups, the policy-implementing agents, the available budget, the available information on the initially expected energy-savings impact, and the cost effectiveness of the instrument
26.039	Distinctions between price-based or quantity-based policy instruments
26.040	Different renewable energy sources and support design elements
26.041	Variable scenarios in the short and long run for key policy parameters such as price of certificate, level of obligation, level of sales tax and the level of penalty
26.042	Fixed-price policies and endogenous price policies
26.043	Technology specific hurdle rates reflecting market barriers, consumer preferences and risk factors limiting purchase of new energy technologies
26.044	Conditions for implementation and proper utilization of saving options (technology equipment availability, familiarity with the policy, overcoming barriers, motivation to invest)
26.045	Specific implementation of policy instruments with regard to their funding
27.002	State-level mandates concerning climate change (national)
27.010	Vehicular emission
27.018	Long-term goals and detailed targets for GHG emissions
27.021	Emissions trend forecast
27.025	Public awareness, education, and participation
27.026	Inter-organizational coordination procedures

Table A4. *Cont.*

Code	Variable
27.027	GHG emissions reduction fee
27.028	Establish a carbon tax
27.029	Disaster resistant land use and building code
27.030	Mixed use and compact development
27.031	Infill development and reuse of remediated brownfield sites
27.032	Green building and green infrastructure
27.033	Low-impact design for impervious surface
27.034	Control of urban service/growth boundaries
27.035	Alternative transportation strategies
27.036	Transit-oriented development and corridor improvements
27.037	Parking standards adjustment
27.038	Pedestrian/resident friendly, bicycle friendly, transit-oriented community design
27.039	Renewable energy and solar energy
27.040	Energy efficiency and energy stars
27.041	Landfill methane capture strategies
27.042	Zero waste reduction and high recycling strategy
27.043	Creation of conservation zones or protected areas
27.044	Watershed-based and ecosystem-based land management
27.045	Vegetation protection
27.048	Identify roles and responsibilities among sectors and stakeholders
27.050	Overall energy efficiency gains
27.053	Energy efficiency gains
27.054	Energy intensity (industry, manufacturing, chemicals, European Union structure)
27.055	Specific consumption (steel, cement, paper)
27.057	Consumption per unit of traffic (road car equivalent, freight, air)
27.058	Car efficiency (fleet average, new cars)
27.060	Energy efficiency gains

Table A5. Variables excluded to avoid the presence of dependent variables in the systematic literature review process.

Code	Variable
1.002	Favorable public opinion
1.006	Social appeal
3.001	Energy statistics (statistical figures)
3.007	Structure, final demand (Reference final demand per subsector)
3.008	Savings, final demand (Actual final heat/electricity demand)
3.009	Conversion efficiency end-users (Actual efficiency boilers, etc.)
4.015	Institutional factors
4.019	Mode of governance
7.009	Production of goods per type (piece, kg or m^2)
7.010	Cultivated land area
7.011	Raising of cattle, pigs and chickens
7.012	Volume index of production
8.001	Gross domestic product (GDP)
11.004	Means—Policy process
12.014	Penetration of energy carriers into useful thermal energy demand of the different sectors
12.018	Average intracity distance traveled per person per day
12.019	Average load factor of intracity passenger transportation mode(s)
12.022	Average intercity distance traveled per person per day
12.024	Average intercity distance driven per car per year
12.025	Average load factor of intercity transportation mode(s)
12.026	Share of car type(s) in the intercity transportation mode(s)

Table A5. *Cont.*

Code	Variable
12.035	Share of urban dwellings with air conditioning, by dwelling type
12.040	Electricity penetration for appliances in urban households
12.042	Penetration of various energy forms into space heating in urban households
12.043	Efficiency of various fuels use, relative to that of electricity use, for space heating in urban households
12.044	Coefficient of performance of heat pumps for space heating in urban households
12.045	Penetration of various energy forms into water heating in urban households
12.046	Efficiency of various fuels use, relative to that of electricity use, for water heating in urban households
12.047	Coefficient of performance of heat pumps for water heating in urban households
12.048	Approximate share of water heating demand in urban households that can be met with solar installations
12.049	Penetration of various energy forms into cooking in urban households
12.050	Efficiency of various fuels use, relative to that of electricity, for cooking in urban households
12.051	Approximate share of cooking demand in urban households that can be met with solar installations
12.052	Share of air conditioning demand of urban households that can be met with electricity
12.053	Coefficient of performance of electric air conditioning in urban households
12.054	Coefficient of performance of non-electric air conditioning in urban households
12.061	Share of rural dwellings with air conditioning, by dwelling type
12.066	Electricity penetration for appliances in rural households
12.068	Penetration of various energy forms into space heating in rural households
12.069	Efficiency of various fuels use, relative to that of electricity use, for space heating in rural households
12.070	Coefficient of performance of heat pumps for space heating in rural households
12.071	Penetration of various energy forms into water heating in rural households
12.072	Efficiency of various fuels use, relative to that of electricity use, for water heating in rural households
12.073	Coefficient of performance of heat pumps for water heating in rural households
12.074	Approximate share of water heating demand in rural households that can be met with solar installations
12.075	Penetration of various energy forms into cooking in rural households
12.076	Efficiency of various fuels use, relative to that of electricity, for cooking in rural households
12.077	Approximate share of cooking demand in rural households that can be met with solar installations
12.078	Share of air conditioning demand of rural households that can be met with electricity
12.079	Coefficient of performance of electric air conditioning in rural households
12.080	Coefficient of performance of non-electric air conditioning in rural households
12.081	Share of service sector in the total active labor force
12.082	Average floor area per employee in the Service sector
12.083	Share of Service sector floor area requiring space heating
12.084	Share of Service sector floor area requiring space heating that is actually heated
12.085	Specific space heat requirements of Service sector floor area
12.086	Share of air-conditioned Service sector floor area
12.087	Specific cooling requirements in the Service sector
12.088	Energy intensity of motor fuel use per Services' subsector
12.089	Energy intensity of electricity specific uses per Services' subsector
12.090	Energy intensity of thermal uses (except space heating) per Services' subsector
12.091	Penetration of various energy forms into space heating in Service sector
12.092	Penetration of various energy forms into other thermal uses in the Service sector
12.093	Efficiency of various fuels use, relative to that of electricity use, for thermal uses in Service sector
12.094	Coefficient of performance of heat pumps in space heating in Service sector
12.095	Share of low-rise buildings in the total Service sector floor area
12.096	Approximate share of thermal uses in the Service sector that can be met by solar installations
12.097	Share of air-conditioning that can be met with electricity
12.098	Coefficient of performance of electric air conditioning in Service sector
12.099	Coefficient of performance of non-electric air conditioning in Service sector

Table A5. *Cont.*

Code	Variable
13.009	Energy per appliance
13.013	Tons-kilometers
13.014	Share of total ton-kilometers by mode
13.020	Energy per value-added by subsector
14.001	Population growth
14.004	Share of the population outside the community
15.004	Irrigation development
15.005	Mechanization
15.008	Industrialization infrastructure
15.011	Poverty
15.014	State policy
15.022	Mass-consuming view of nature
16.007	Measures and policies already promulgated
16.012	Implementation experiences
17.003	Average income
17.005	Environmental voting
18.003	Legal structure and status
18.004	Spending power
18.005	Control over income
20.003	Age of occupants
20.005	Dwelling type
21.004	Price development
21.005	Sales data
23.001	Attitude factors
23.002	Economic factors
23.003	Social factors (impact on agents' attitude)
24.010	Average energy/value-added per subsector
25.004	Motorization rate
25.005	Job intensity
26.006	Different renewable energy sources penetration levels
26.010	Discount rate for investments
26.013	Socio-demographic and lifestyle trends
26.015	Political context in which policy instruments are imposed
26.046	Transaction costs
26.047	Stability and credibility in policy regime
27.001	Political will
27.003	Wealth
27.005	Population density
27.051	Final intensity
27.052	Primary intensity

References

1. Meeus, L.; Oliveira Fernandes, E.; Delarure, E.; Azevedo, I.; Leal, V.M.S.; Glachant, J.M. *Smart Cities Initiative: How to Foster A Quick Transition towards Local Sustainable Energy Systems*; THINK Project, EU 7th Framework Programme; European University Institute: Florence, Italy, 2011.
2. Azevedo, I.; Leal, V.M.S. Methodologies for the evaluation of local climate change mitigation actions: A review. *Renew. Sustain. Energy Rev.* **2017**, *79*, 681–690. [CrossRef]
3. Millard-Ball, A. Do city climate plans reduce emissions? *J. Urban Econ.* **2012**, *71*, 289–311. [CrossRef]
4. Millard-Ball, A. The Limits to Planning: Causal Impacts of City Climate Action Plans. *J. Plan. Educ. Res.* **2012**, *33*, 5–19. [CrossRef]
5. Azevedo, I.; Horta, I.; Leal, V.M.S. Analysis of the relationship between local climate change mitigation actions and greenhouse gas emissions-Empirical insights. *Energy Policy* **2017**, *111*, 204–2013. [CrossRef]
6. Pullin, A.S.; Knight, T.M. Data credibility: A perspective from systematic reviews in environmental management. *New Dir. Eval.* **2009**, *122*, 65–74. [CrossRef]

7. Auld, G.; Mallett, A.; Burlica, B.; Nolan-Poupart, F.; Slater, R. Evaluating the effects of policy innovations: Lessons from a systematic review of policies promoting low-carbon technology. *Glob. Environ. Chang.* **2014**, *29*, 444–458. [CrossRef]
8. Boonekamp, P.G.M. Energy and emission monitoring for policy use-Trend analysis with reconstructed energy balances. *Energy Policy* **2004**, *32*, 969–988. [CrossRef]
9. Boonekamp, P.G.M. *Improved Methods to Evaluate Realised Energy Savings*; Proefschrift Universiteit Utrecht: Utrecht, The Netherlands, 2005.
10. Broto, V.C.; Bulkeley, H. A survey of urban climate change experiments in 100 cities. *Glob. Environ. Chang.* **2013**, *23*, 92–102. [CrossRef]
11. Covenant of Mayors Office and Joint Research Centre of the European Commission. *Reporting Guidelines on Sustainable Energy Action Plans*; Publications Office of the European Union: Luxembourg, 2010.
12. Corfee-Morlot, J.; Kamal-Chaoui, L.; Donovan, M.G.; Cochran, I.; Robert, A.; Teasdale, P.J. Cities, Climate Change and Multilevel Governance. *OECD Environ. Work. Pap.* **2009**. [CrossRef]
13. Farla, J.C.M.; Blok, K. The use of physical indicators for the monitoring of energy intensity development in the Netherlands, 1980–1995. *Energy* **2000**, *25*, 609–638. [CrossRef]
14. Fortes, P.; Alvarenga, A.; Seixas, J.; Rodrigues, S. Long-term energy scenarios: Bridging the gap between socio-economic storylines and energy modeling. *Technol. Forecast. Soc. Chang.* **2015**, *91*, 161–178. [CrossRef]
15. Fortes, P.; Pereira, R.; Pereira, A.M.; Seixas, J. Integrated technological-economic modelling platform for energy and climate policy analysis. *Energy* **2014**, *73*, 716–730. [CrossRef]
16. Gustavsson, E.; Elander, I.; Lundmark, M. Multilevel governance, Networking cities, and the geography of climate-change mitigation: Two Swedish examples. *Environ. Plan. C Gov. Policy* **2009**, *27*, 59–74. [CrossRef]
17. Gysen, J. The Modus Narrandi: A Methodology for Evaluating Effects of Environmental Policy. *Evaluation* **2006**, *12*, 95–118. [CrossRef]
18. International Atomic Energy Agency (IAEA). Model for Analysis of Energy Demand (MAED-2)-User's Manual. In *Computer Manual Series*; IAEA: Vienna, Austria, 2006.
19. International Energy Agency (IEA). Energy Efficiency Indicators: Essentials for Policy Making. In *Energy Efficiency Indicators*; IEA: Paris, France, 2014.
20. Jovanovic, M.; Afgan, N.; Bakic, V. An analytical method for the measurement of energy system sustainability in urban areas. *Energy* **2010**, *35*, 3909–3920. [CrossRef]
21. Kasperson, J.X.; Kasperson, R.E.; Turner, B.L. *Regions at Risk: Comparison of Threatened Environments*; United Nations University Press: Tokyo, Japan, 1995.
22. Lin, J.; Cao, B.; Cui, S.; Wang, W.; Bai, X. Evaluating the effectiveness of urban energy conservation and GHG mitigation measures: The case of Xiamen city, China. *Energy Policy* **2010**, *38*, 5123–5132. [CrossRef]
23. Morlet, C.; Keirstead, J. A comparative analysis of urban energy governance in four European cities. *Energy Policy* **2013**, *61*, 852–863. [CrossRef]
24. Murphy, L.; Meijer, F.; Visscher, H. A qualitative evaluation of policy instruments used to improve energy performance of existing private dwellings in the Netherlands. *Energy Policy* **2012**, *45*, 459–468. [CrossRef]
25. Nässén, J. Determinants of greenhouse gas emissions from Swedish private consumption: Time-series and cross-sectional analyses. *Energy* **2014**, *66*, 98–106. [CrossRef]
26. Neij, L.; Åstrand, K. Outcome indicators for the evaluation of energy policy instruments and technical change. *Energy Policy* **2006**, *34*, 2662–2676. [CrossRef]
27. Praznik, M.; Butala, V.; Zbašnik Senegačnik, M. Simplified evaluation method for energy efficiency in single-family houses using key quality parameters. *Energy Build.* **2013**, *67*, 489–499. [CrossRef]
28. Rai, V.; Robinson, S.A. Agent-based modeling of energy technology adoption: Empirical integration of social, behavioral, economic, and environmental factors. *Environ. Model. Softw.* **2015**, *70*, 163–177. [CrossRef]
29. Schipper, L.; Unander, F.; Murtishaw, S.; Ting, M. Indicators of Energy Use and Carbon Emissions: Explaining the Energy Economy Link. *Annu. Rev. Energy Environ.* **2001**, *26*, 49–81. [CrossRef]
30. Sobrino, N.; Monzon, A. The impact of the economic crisis and policy actions on GHG emissions from road transport in Spain. *Energy Policy* **2014**, *74*, 486–498. [CrossRef]
31. Spyridaki, N.A.; Flamos, A. A paper trail of evaluation approaches to energy and climate policy interactions. *Renew. Sustain. Energy Rev.* **2014**, *40*, 1090–1107. [CrossRef]

32. Tang, Z.; Brody, S.D.; Quinn, C.; Chang, L.; Wei, T. Moving from agenda to action: Evaluating local climate change action plans. *J. Environ. Plan. Manag.* **2010**, *53*, 41–62. [CrossRef]
33. Wilbanks, T.J.; Kates, R.W. Global change in local places: How scale matters. *Clim. Chang.* **1999**, *43*, 601–628. [CrossRef]

© 2020 by the authors. Licensee MDPI, Basel, Switzerland. This article is an open access article distributed under the terms and conditions of the Creative Commons Attribution (CC BY) license (http://creativecommons.org/licenses/by/4.0/).

Article

Wind Power Cogeneration to Reduce Peak Electricity Demand in Mexican States Along the Gulf of Mexico

Quetzalcoatl Hernandez-Escobedo [1], Javier Garrido [1], Fernando Rueda-Martinez [1], Gerardo Alcalá [2] and Alberto-Jesus Perea-Moreno [3,*]

1 Faculty of Engineering, Campus Coatzacoalcos, Universidad Veracruzana Mexico, Coatzacoalcos, Veracruz 96535, Mexico; qhernandez@uv.mx (Q.H.-E.); jgarrido@uv.mx (J.G.); frueda@uv.mx (F.R.-M.)

2 Centro de Investigación en Recursos Energéticos y Sustentables, Universidad Veracruzana Mexico; Coatzacoalcos, Veracruz 96535, Mexico; galcala@uv.mx

3 Departamento de Física Aplicada, Universidad de Córdoba, ceiA3, Campus de Rabanales, 14071 Córdoba, Spain

* Correspondence: aperea@uco.es; Tel.: +34-957-212-633

Received: 14 May 2019; Accepted: 10 June 2019; Published: 18 June 2019

Abstract: The Energetic Transition Law in Mexico has established that in the next years, the country has to produce at least 35% of its energy from clean sources in 2024. Based on this, a proposal in this study is the cogeneration between the principal thermal power plants along the Mexican states of the Gulf of Mexico with modeled wind farms near to these thermal plants with the objective to reduce peak electricity demand. These microscale models were done with hourly MERRA-2 data that included wind speed, wind direction, temperature, and atmospheric pressure with records from 1980–2018 and taking into account roughness, orography, and climatology of the site. Wind speed daily profile for each model was compared to electricity demand trajectory, and it was seen that wind speed has a peak at the same time. The amount of power delivered to the electric grid with this cogeneration in Rio Bravo and Altamira (Northeast region) is 2657.02 MW and for Tuxpan and Dos Bocas from the Eastern region is 3196.18 MW. This implies a reduction at the peak demand. In the Northeast region, the power demand at the peak is 8000 MW, and for Eastern region 7200 MW. If wind farms and thermal power plants work at the same time in Northeast and Eastern regions, the amount of power delivered by other sources of energy at this moment will be 5342.98 MW and 4003.82 MW, respectively.

Keywords: wind farm; thermal power plants; peak electricity demand; Gulf of Mexico

1. Introduction

One of the world's biggest problems is that population always is increasing and some countries have over dependence on energy generation. In the world, primary energy consumption growth averaged 2.2% in 2017, up from 1.2% last year and the fastest since 2013. This compares with the 10-year average of 1.7% per year [1]. In 2016, generation from combustible fuels accounted for 67.3% of total world gross electricity production (of which: 65.1% from fossil fuels; 2.3% from biofuels and waste), hydroelectric plants (including pumped storage) provided 16.6%; nuclear plants 10.4%; geothermal, solar, wind, tide, and other sources 5.6%; and biofuels and waste made up the remaining 2.3% [2]. It is known that energy supply side and demand side is essential to the correct usage of energy [3], and this is important to nations because it depends on the effectiveness of its power systems and the availability of power supply at the load centers [4]. It is important to define the concepts of load and electricity demand, on the one hand to understanding load of an electric power distribution as the final stage of the system that converts electric energy in another form of energy; respecting electricity demand is known as the electric power related on a period of time, that uses load to work.

In Mexico, electricity demand is considered decisive in social and economic development and consequently in the improvement of the economic conditions. The maximum demand for electrical energy occurs mainly during the summer season, between May and September, when the heat is more severe in the north and southeast of the country. As a result, the use of ventilation equipment is increased to maintain the temperature at more comfortable levels and the values necessary for the proper functioning of certain appliances are regulated. In 2017, the energy independence index, which shows the relationship between production and national energy consumption, was equivalent to 0.76. This result implies that the amount of energy produced in the country was 24.0% less than that which was made available in the various consumer activities in the national territory. In the course of the last ten years, this indicator decreased in an average annual rate of 5.0% [5]. An alternative will be the use of renewable energy, the Transition Energetic Law establish a renewable energy minimal participation in electricity generation of 25%, 30% and 35% in 2018, 2021, and 2024, respectively [6]. In the energetic mix, renewable energies have a participation of 14%, the technologies most used are hydraulic, and wind energy contributed with 3% or 38.23 PJ in 2017 [7].

Peak electricity demand is a global policy concern which creates transmission constraints and congestion, and raises the cost of electricity for all end-users [8]. Also, a considerable investment is required to upgrade electricity distribution and transmission infrastructure, and build generation plants to provide power during peak demand periods [9]. This is important because usually service suppliers charge a higher price for services at peak-time than for off-peak time to compensate for the costly electricity generation at peak hours [10]. Thus, if peak demand is reduced, electric system will be benefited. In addition to these benefits a study done by the authors of Reference [11] determine that it can eliminate the need to install expensive extra generation capacity such as combustion turbines for peak hours which are less than a hundred hours a year. With this information regions will be considered where demand complies with the conditions to have the necessary hours. Another benefit is for consumers, because if they know the period of time of peak demand and decrease their consumption, they can avoid higher electricity prices. An option to reduce peak demand is installing peak plants but the economic benefit does not justify the investment. Usually these peak plants use fossil fuel and the result is more generation of greenhouse gases [12].

The use of wind energy to support the electric demand mainly during its peak period is an alternative to reduce fossil fuel combustion and greenhouse gas emissions. Wind is intermittent and uncertain, which represents a problem in power systems [13]. A study done by the authors of Reference [14] established that due to the unpredictable nature of wind energy and non-coincidence between wind units output power and demand peak load, wind units are deemed as an unreliable source of energy. In this study a stochastic mathematical model was developed for the optimal allocation of energy storage units in active distribution networks in order to reduce wind power spillage and load curtailment while managing congestion and voltage deviation.

In New York State the intraregional effects were illustrated by quantifying the net load, net load ramping, operating reserve, and regulation requirements, the study found out that only at wind capacities exceeding 100% of the average statewide load does the wind-generated electricity meet significant portions of the distant demands [15].

An improved energy hub system combined with cooling, heating, and power integrated with photovoltaic and wind turbine, demonstrated that operation cost and carbon emission were reduced by 3.9% and 2.26%, respectively [16]. In Canada a battery storage system to minimize the energy drawn from the public grid was proposed This storage system was charged from wind turbines and the energy stores were discharged to a park when the wind park power dropped below 0 kW and the storage system was able to offset 17.2 MWh but the financial gain was insufficient to offset the net energy losses in the storage system [17].

About cogeneration, in China Reference [18] established the relationship between a wind farm and a thermal power plant, in this study the difference between wind power profits and thermal power is only analyzed in terms of cost differences, and no comparison of other data is involved.

According to this, the specific example of Gansu Province not only shows the difference in earnings before and after the peak-to-peak adjustment of wind power, but also indicates that the large-scale auxiliary adjustment of wind power can ensure that thermal power generators can participate in deep-peaking with reasonable returns; Reference [19] analyses and compares several existing kinds of distributed energy storage for improving wind power integration. Then considering a mass of cogeneration units in north China, control strategy for wind power integration is proposed to reduce the peak regulation capacity for wind power integration. As a result, the peak–valley load difference can be equivalently reduced to relieve peaking pressure for wind power integration.

In Northern China a proposal were made suggesting that if the energy carrier for part of the end users space heating is switched from heating water to electricity (e.g., electric heat pumps can provide space heating in the domestic sector), the ratio of electricity to heating water load should be adjusted to optimize the power dispatch between cogeneration units and wind turbines, resulting in fuel conservation, they found out that with the existing infrastructures are made full use of, and no additional ones are required [20]. Bexten et al. [21] investigate a system configuration, which incorporates a heat-driven industrial gas turbine interacting with a wind farm providing volatile renewable power generation. The study quantifies the impact of selected system design parameters on the quality of local wind power system integration that can be achieved with a specific set of parameters. Results show that the investigated system configuration has the ability to significantly increase the level of local wind power integration.

In this study a methodology is proposed, which consists in locating thermal power plants along the five states of the Gulf of Mexico (Tamaulipas, Veracruz, Tabasco, Campeche, and Yucatan) and evaluating wind resources next to these plants with the objective to determine the power output generated and their contribution to reduce the peak electricity demand. For the geographic position of Mexican states along the Gulf of Mexico, see Figure 1.

Figure 1. Mexican states along the Gulf of Mexico.

2. Material and Methods

2.1. Data

Modern-Era Retrospective analysis for Research and Applications, Version 2 (MERRA-2) provides re-analysis data such as: wind speed and wind direction at 50 m height, temperature at 2 m and 10 m, and atmospheric pressure with records from 1980–2018. The resolution of every point downloaded is 50 km covering all the Mexican states along the Gulf of Mexico including offshore points, which can all be observed in Figure 2. Several studies have used MERRA-2 to determine wind characteristics, as did Reference [22] in the Central Californian Coast, assessing offshore conditions. A study about wind and rainfall areas of tropical cyclones making landfall over South Korea was examined for the period 1998–2013 using MERRA-2 data. It was determined that composite analyses of the cases of strong and weak vertical wind shear confirm that the increase of rainfall area is related to the asymmetric convection (rising/sinking motion in the downshear-left/upshear-right side) induced by the vertical wind shear [23]. An investigation on prediction of wind power because of climate change was made using MERRA-2 among other re-analysis data [24]. As can be seen, MERRA-2 data have been used with excellent results on wind assessment.

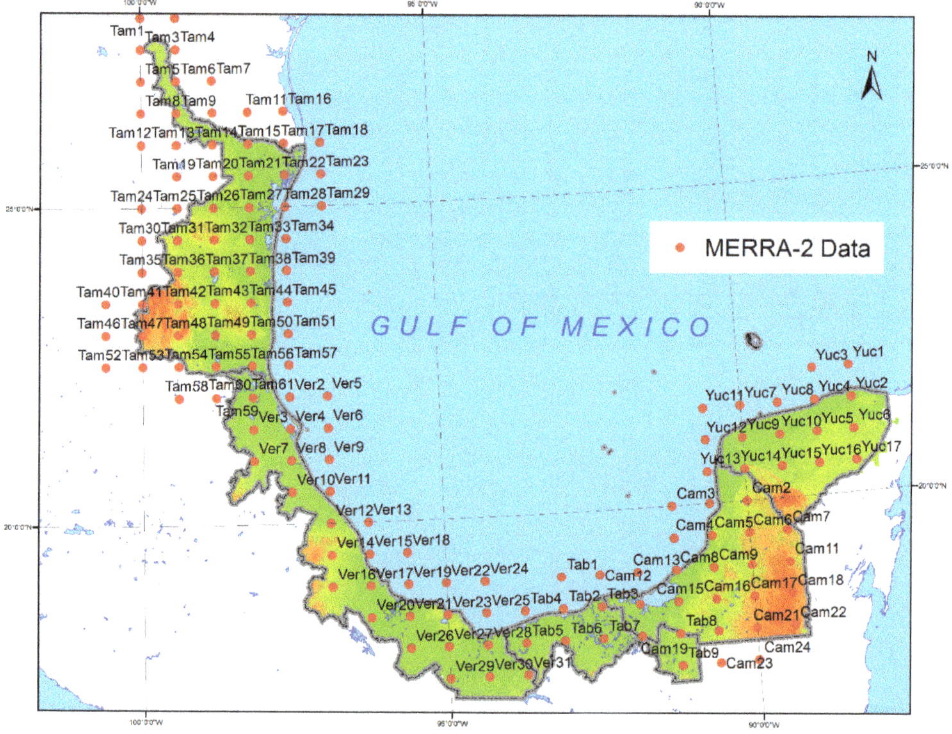

Figure 2. Modern-Era Retrospective analysis for Research and Applications, Version 2 (MERRA-2) data position.

In total, 142 MERRA-2 points, 61 in Tamaulipas, 31 in Veracruz, 9 in Tabasco, 24 in Campeche and 17 in Yucatan.

2.2. Thermal Power Plants

There are 22 thermal power plants in Mexico that generate electricity. These plants work on a combined cycle with gas turbines. This generation contributes 55.6% of all electricity in the country [25]. Along the Gulf of Mexico, the most important plants are located as follows: two in the state of Veracruz, one called Adolfo Lopez Mateos with capacity of 2263 MW and Dos Bocas with capacity 350 MW. In the state of Tamaulipas there are two, Rio Bravo and Altamira, with capacity of 800 MW and 1039 MW respectively [26]. Figure 3 shows the geographic position of these thermal power plants. Unfortunately, there are no thermal plants located in the states of Tabasco, Campeche, and Yucatan.

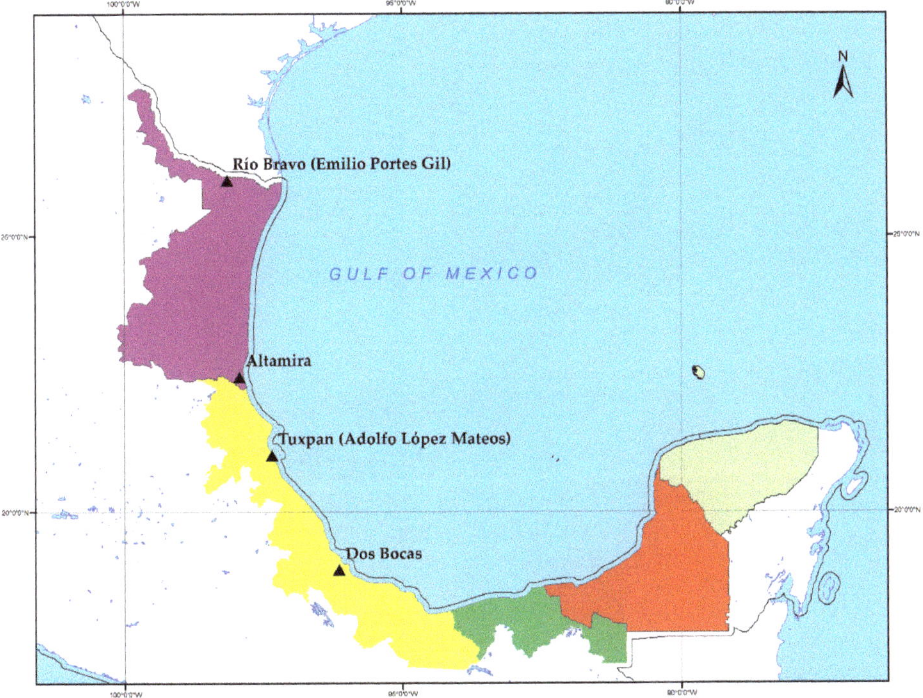

Figure 3. Thermal power plants.

2.3. Wind Assessment

Wind characterization is essential to determine its power at a given site. Its assessment depends mainly on the orography, roughness, and generalized climatology.

Statistically it can be characterized as a probability density function (PDF) that can be interpreted as the probability that the random variable x lies in a differential range, dx, about a value x is $f(x)dx$. More specific statements about the probability that the unsteady variable x, lies in a particular range $a \leq x \geq b$, this expression is given by Equation (1).

$$P = (a \leq x \leq b) = \int_a^b f(x)dx, \tag{1}$$

Another PDF used in wind assessment is the Weibull distribution for the wind speed (u) that has two parameters: shape parameter (k) and scale parameter (c). This is expressed in Equation (2):

$$f(u) = \frac{k}{c}\left(\frac{u}{c}\right)^{k-1} e^{-(\frac{u}{c})^k}, \tag{2}$$

The power in the wind is the product of mass flow rate through turbine blades ρUA, and the kinetic energy per unit mass in the wind, $U^2/2$. The average power in the incoming wind is presented in Equation (3) [27].

$$P = \frac{1}{2}\rho A U^3 C_P, \tag{3}$$

where ρ is the density of air, C_P is the power coefficient, A is the rotor swept area, U is the wind speed and P is the wind turbine power output.

2.4. Peak Electrical Demand

In Mexico the National Energy Control Center (NECC) divided the country in seven regions. This division is called National Interconnected System [28], and is based in two electric regions: Northeast and Eastern [29]. These regions can be seen in Figure 4.

Figure 4. Regions analyzed.

NECC shows information about demand forecasting, net demand, and total demand. It uses three methodologies to calculate demand forecasting: moving average that is optimal for random or leveled demand patterns where the impact of historical irregular elements is to be eliminated through an approach of periods of up to 7 days previously. This methodology is the most used by NECC and is given by Equation (4)

$$\hat{X}_t = \frac{\sum_{i=1}^{n} x_{t-1}}{n}, \tag{4}$$

where $\hat{X}_t$ is the energy demand average in period t; X_{t-1} is the real demand for the periods prior to t and the number of observations n.

The second method is weighted moving average, a variation of moving average. While in moving average all data has the same weight, in weighted moving average each data is assigned with a specific weight, always considering that total sum will be 100% and is expressed as Equation (5):

$$\hat{X}_t = \sum_{i=1}^{n} c_i x_{t-1}, \tag{5}$$

where $\hat{X}_t$ is the energy demand average in period t; c_i weighted factor; x_{t-1} is the real demand for the periods prior to t and the number of observations n.

The third method is called multiple linear regression, where both independent and dependent variables working with the independent variable variation to forecasting the dependent variable. Multiple linear regression is given by Equation (6):

$$D_F = \beta_0 + \beta_1 x_1 + \cdots + \beta_n x_n + \varepsilon, \tag{6}$$

where D_F is Demand Forecasted; $x_1, \ldots\ x_n$ independent variables; $\beta_0 \ldots\ \beta_n$ coefficients which are calculated by least squares; n number of data and ε is a random error term.

Several studies about peak demand where demand consumers participate have established two typical types of programs. The first one is an incentive-based program the end users are encouraged by the incentive payments to reduce their load demands, and thus to help improve the feasibility and stability of the power system [30]. This program usually includes direct load control and demand bidding [31]. The second one is price-based, and in this case final consumers are encouraged to configure their energy usage patterns according to electricity demand curve. These price-based programs consist in observing critical peaking pricing, time of use pricing, and time pricing [32,33].

Figure 5 presents two demand curves, each one by region. These demand curves have been developed with the aim to estimate the current hourly electricity demand profiles with a temporal resolution of 1 h based in 2018.

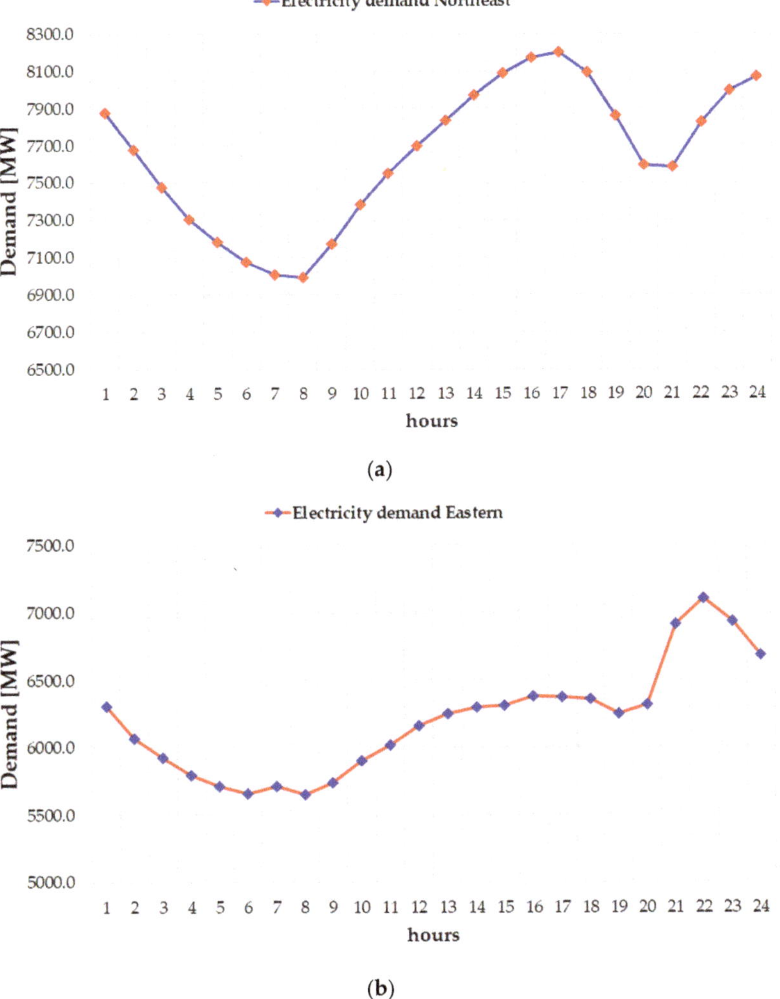

Figure 5. (**a**) Northeast electricity demand and (**b**) Eastern electricity demand.

As seen in Figure 5, the behavior of electric demand varies among regions, Figure 5a at Northeast the peak appears in range of 16:00–18:00 h and has another peak during 22:00–24:00 h. This is because the consumption increases in the first peak due to domestic use and the second electricity peak is due to industrial sector and in Figure 5b the Eastern region presents a peak demand during 21:00–23:00 h, mainly for industrial activities.

These demand curves will be used and compared to wind daily profile to consider it as wind power.

3. Results and Discussion

Along the Gulf of Mexico 142 MERRA-2 points have been assessed with wind speed information. 4 sites around thermal power plants have been chosen and the amount of power output generated has been calculated to reduce peak electricity demand.

3.1. Wind Resource Assessment

Wind has been converted as resource along the Gulf of Mexico; this means that mainly both roughness and orography have been taken into account.

3.1.1. Rio Bravo, Altamira, Tuxpan, and Dos Bocas

Four MERRA-2 points were located near thermal power plants, their geographic information is presented in Table 1.

Table 1. Geographic information for Modern-Era Retrospective analysis for Research and Applications, Version 2 (MERRA-2) and thermal power plants.

Thermal Plant	Lat [N]	Lon [W]	MERRA-2	Lat [N]	Lon [W]
Rio Bravo	26.0033°	98.1335°	Tam15	26°	98.125°
Altamira	22.4945°	97.9029°	Tam56	22.5°	98°
Tuxpan	21.0199°	97.3448°	Ver	21°	97.5°
Dos Bocas	19.0840°	96.1491°	Ver19	19°	96.25°

In Appendix A Tables A1–A4 hourly mean wind speed are presented by month for MERRA-2 points. In each one of them wind speed behavior can be appreciated.

A typical wind year can be compared to electricity demand to identify its trajectory, so that it could be seen if in the peak demand there existed a peak wind. To probe it, Figures 6 and 7 represent both electricity demand and wind speed of Rio Bravo, Altamira, Tuxpan, and Dos Bocas respectively.

As can be observed in Figure 6, wind speed trajectory has an increase in the last hours as the electricity demand does as follows: Figure 6a 6.8 m/s to 7.6 m/s and Figure 6b 5 m/s to 6.5 m/s both during 21:00 h to 24:00 h. Although Figure 6b does not follow the entire trajectory, at the end of the day the last 4 h (21:00–24:00) when the demand begins to grow, wind also does. In Figure 7a 4.5 m/s to 6.5 m/s and Figure 7b 5 m/s to 6 m/s both between 20:00 h to 24:00 h. This could be considered to model a potential source of power.

Determining wind speed features correctly means that the model that will be designed will have more accuracy. In Table 2, statistical information at the points studied is presented.

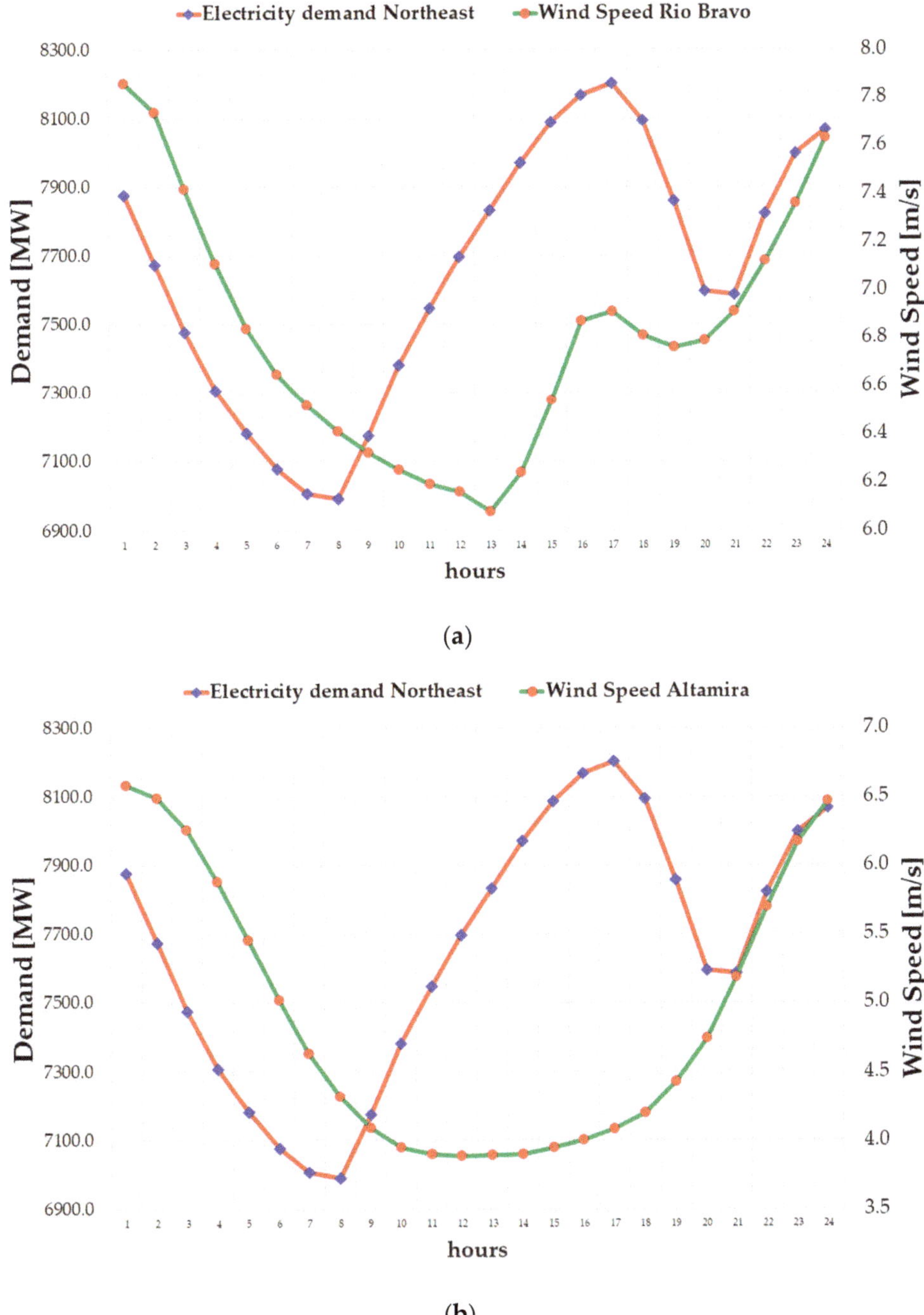

Figure 6. (**a**) Northeast demand compared to wind speed at Rio Bravo and (**b**) Northeast demand compared to wind speed at Altamira.

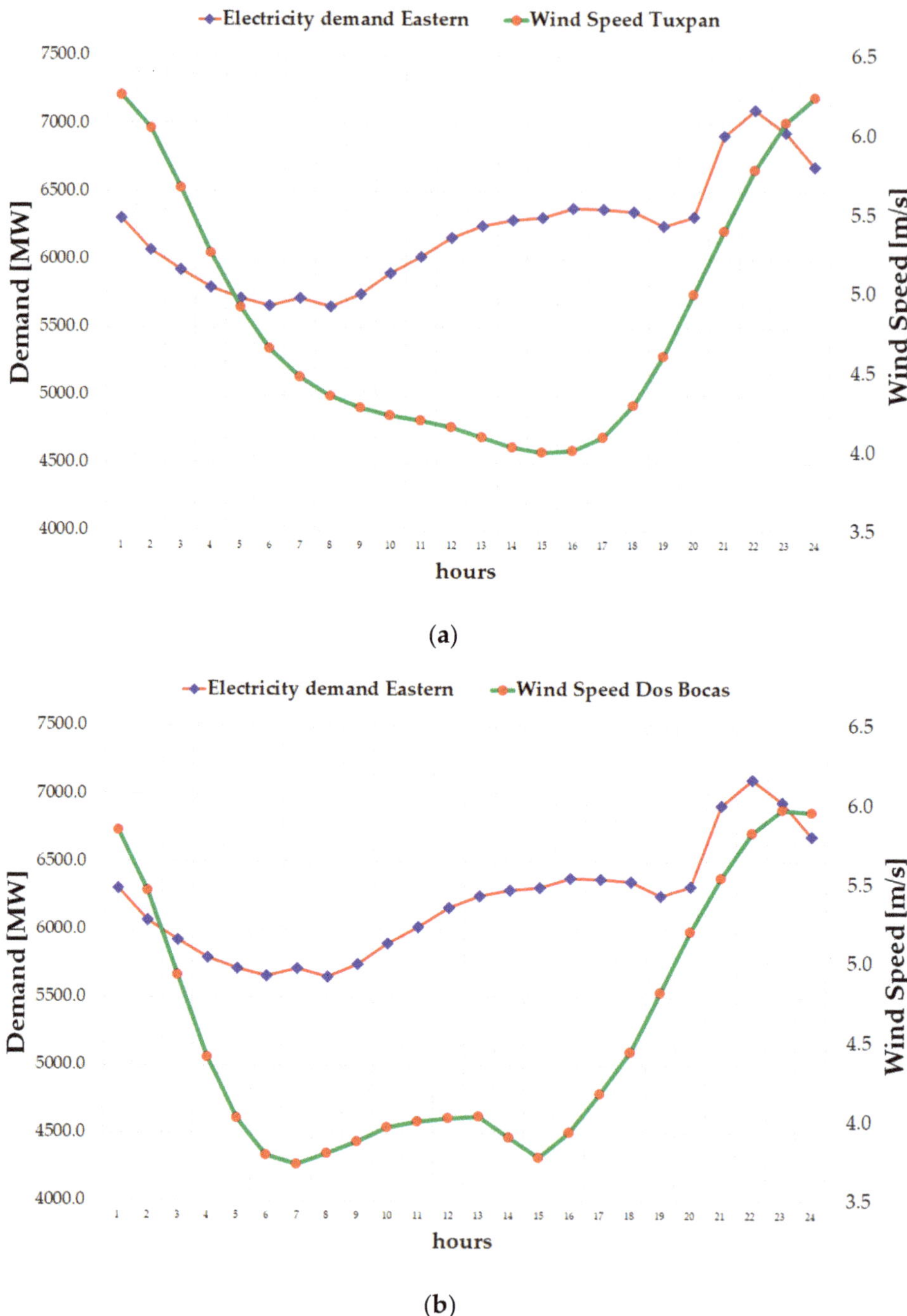

Figure 7. (**a**) Eastern electricity demand compared to wind speed at Tuxpan and (**b**) Eastern electricity demand compared to wind speed at Dos Bocas.

Table 2. Wind resource statistics.

	Variable	Mean	Min	Max
Rio Bravo	Air density [kg/m^3]	1.172	1.169	1.172
	Mean speed [m/s]	6.79	6.76	6.94
	Power density [W/m^2]	265	261	286
	Weibull-c [m/s]	7.6	7.6	7.8
	Weibull-k	2.84	2.78	2.87
Altamira	Air density [kg/m^3]	1.170	1.160	1.171
	Mean speed [m/s]	4.99	4.78	5.46
	Power density [W/m^2]	130	114	170
	Weibull-c [m/s]	5.6	5.4	6.2
	Weibull-k	2.15	2.09	2.21
Tuxpan	Air density [kg/m^3]	1.175	1.173	1.175
	Mean speed [m/s]	5.06	4.75	5.34
	Power density [W/m^2]	139	116	161
	Weibull-c [m/s]	5.7	5.4	6.0
	Weibull-k	2.09	2.03	2.14
Dos Bocas	Air density [kg/m^3]	1.166	1.163	1.168
	Mean speed [m/s]	4.95	4.8	5.19
	Power density [W/m^2]	126	114	144
	Weibull-c [m/s]	5.6	5.4	5.9
	Weibull-k	2.16	2.12	2.2

3.1.2. Microscale Model

One of the critical criteria to place wind turbines is to assess wind resource and determine if the site have features as transmission lines, substations, and roads, among others.

A microscale model has been done on WAsP software including roughness and orography features. In Figures 8 and 9 the four sites are shown microscale modeled. Each one has an array of seven wind turbines placed to avoid wake effect. Wind turbines generators (WTG) are selected according to wind features at the site.

Modeled areas show different types of wind resource, and the resolution is 10 km. The areas of high wind energy potential are better identified. An advantage of microscale modelling is to extrapolate onsite measurements to the prospective turbine sites to modify the background atmospheric conditions and account for topographic induced speedups, turbulence, and wind shear [34]. The x and y axes are represented in UTM scale, in Figure 8a the zone presents a wind power density between 261 W/m^2 to 286 W/m^2, WTG type GEWind 77 were placed to avoid wake effects; Figure 8b presents a zone with 170 W/m^2, level curves show hills in this area and were harnessed to place WTG type BONUS 600 kW along it. In Figure 9a, this zone is in front of the sea, an area with 142 W/m^2, wind turbines VESTAS V82-1650 kW were used; Figure 9b. In Dos Bocas there is a wind power density of 144 W/m^2. Seven WTC type BONUS 600 kW were placed at the top of the hills.

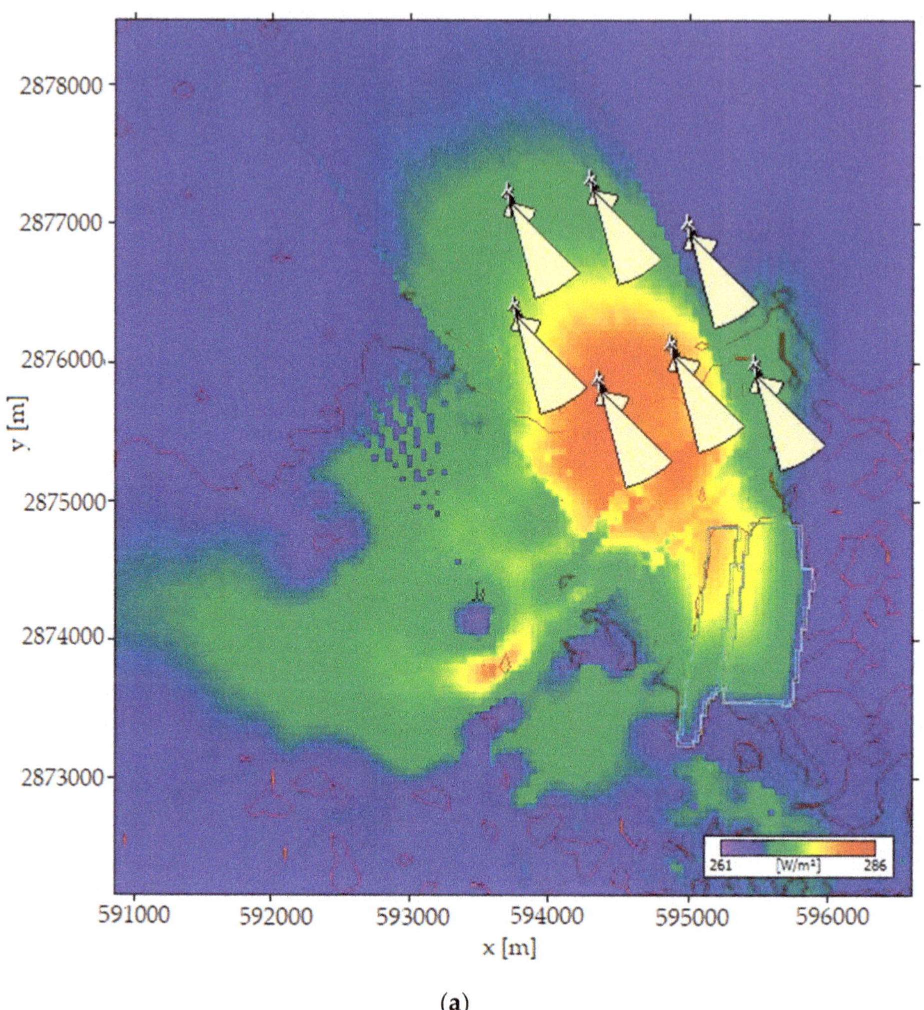

(a)

Figure 8. *Cont.*

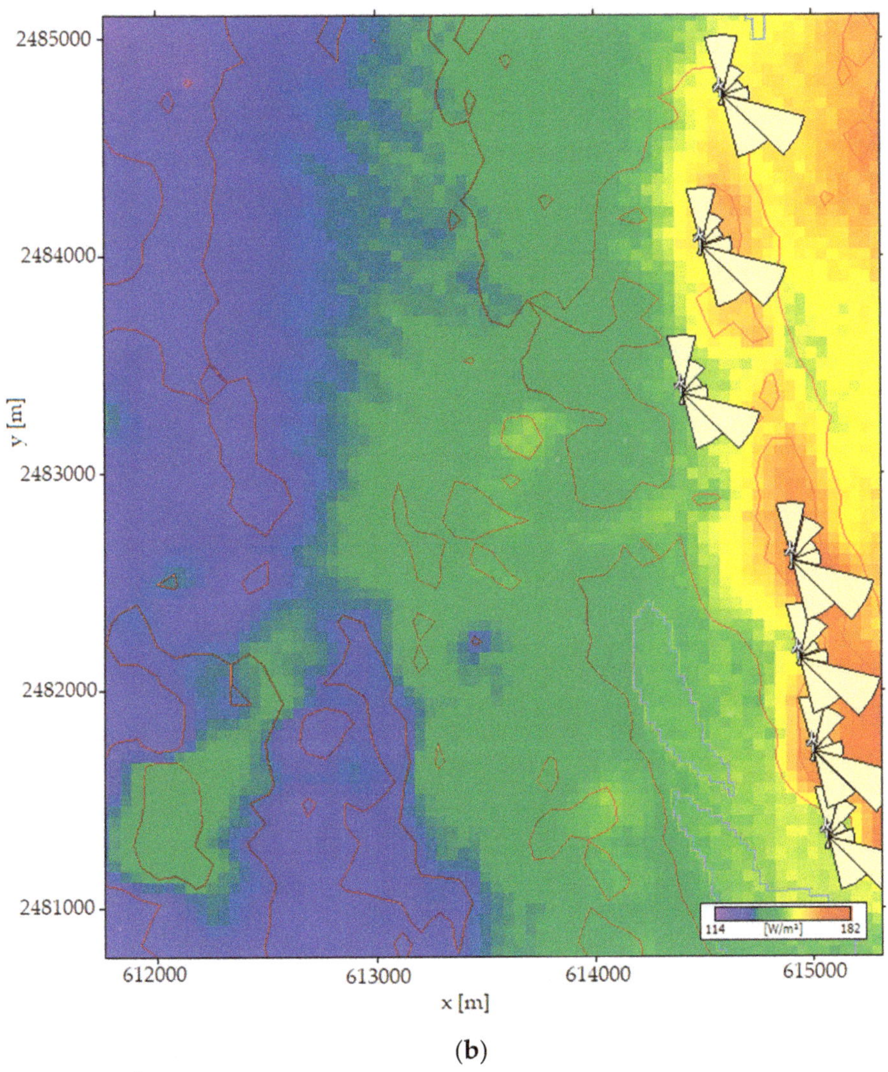

(b)

Figure 8. (**a**) Rio Bravo and (**b**) Altamira.

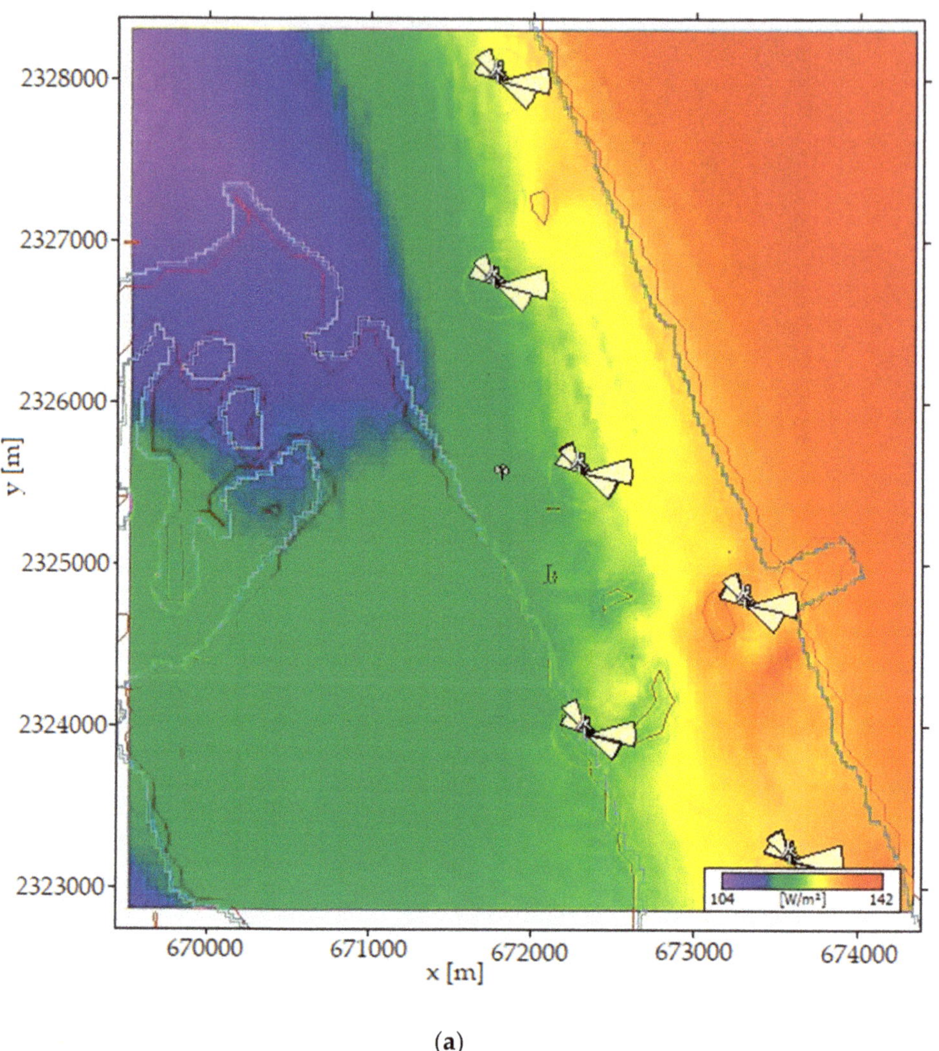

(**a**)

Figure 9. *Cont.*

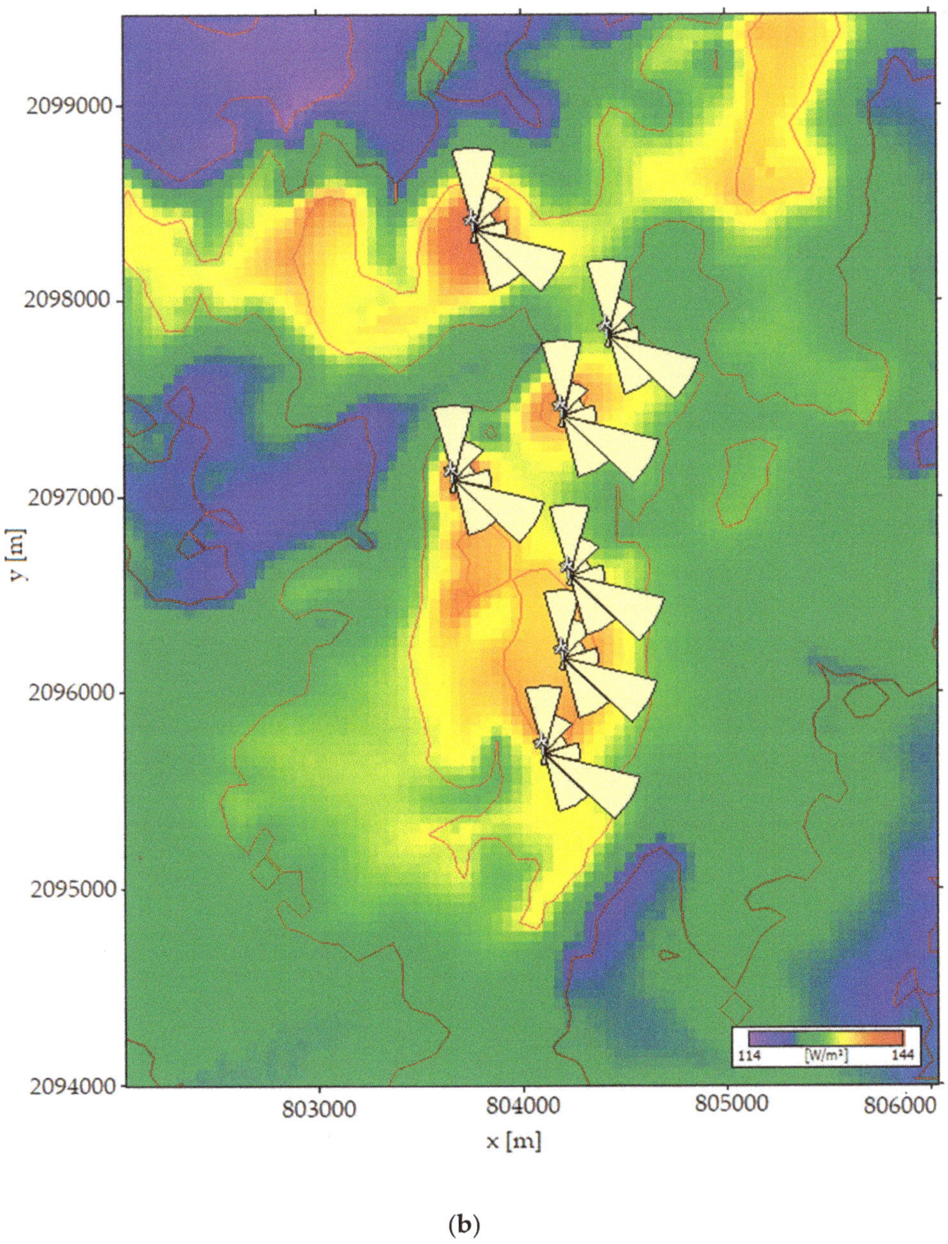

(b)

Figure 9. (a) Tuxpan and (b) Dos Bocas.

3.1.3. Power Output Generation and Electricity Peak Demand

In the northeast region, peak demand occurs twice per day, the first period at 17:00 h and the second one at 24:00 h. These peaks represent a higher price of electricity by consumers. Wind speed at Rio Bravo behaves similarly to electricity demand curve, wind daily profile has two peaks, and these at the same time. Meanwhile, wind speed profile in Altamira only has one peak at the end of the day, even so power can be extracted from the wind. The power contribution by two wind farms placed in Rio Bravo and Altamira in the state of Tamaulipas, respectively, is shown in Table 3.

Table 3. Wind farms power generation along the state of Tamaulipas.

Site	Variable	Total	Mean	Min	Max
Rio Bravo	Power output generated [MW]	9322.9	405.3	0	630
	Total net AEP [GWh]	37.9	5.4	5.4	5.5
	Power density [W/m^2]	-	385	381	389
Altamira	Power output generated [MW]	3684.6	150.1	0	243.1
	Total net AEP [GWh]	6514.2	930.6	880.5	972.8
	Power density [W/m^2]	-	189	179	198

In the state of Veracruz located at the eastern electricity region, wind has been analyzed near two thermal power plants. In this region, electricity demand has a peak that occurs at 22:00 h in both Tuxpan and Dos Bocas, and wind speed trajectory shows that it can contribute delivering electricity to the grid at the same time that peak demand occurs. Table 4 presents the power output generated of two wind farms at the state of Veracruz.

Table 4. Wind farms power generation along the state of Veracruz.

Site	Variable	Total	Mean	Min	Max
Tuxpan	Power output generated [MW]	7589.8	361.4	0	693
	Total net AEP [GWh]	17.6	2.5	2.26	2.63
	Power density [W/m^2]	-	145	133	152
Dos Bocas	Power output generated [MW]	3178.7	143	0	241.9
	Total net AEP [GWh]	6016.76	859.50	847.93	871.05
	Power density [W/m^2]	-	175	172	179

If the period of time and the wind speed are taken into account when peak demand is increasing, it will be able to calculate how much energy could be extracted in each point, see Table 5.

Table 5. Total power generation at peak demand.

Site	Hour@peak Demand	Wind Speed [m/s] @ Peak Demand	Power Generated [MW]	Subtotal [MW]	Thermal Power [MW]	Total
Rio Bravo	15:00	6.5	131.72	719.48	800	1519.48
	16:00	6.9	232.07			
	17:00	7	355.69			
Altamira	22:00	5	18.29	98.54	1039	1137.54
	23:00	6	31.27			
	24:00	7	48.98			
Tuxpan	21:00	5.5	78.35	481.71	2263	2744.71
	22:00	5.9	154.76			
	23:00	6.3	248.60			
Dos Bocas	21:00	5.6	25.26	101.47	350	451.47
	22:00	6	33.63			
	23:00	6.4	42.58			

As shown in Table 5, electric power generated by each wind farm could contribute to reduce peak demand. The electric power generated by region indicate an amount of 2657.02 MW for Northeast and 3196.18 MW for Eastern. This is the power delivered to the electric grid.

Wind resource assessment is fundamental to determine the site where wind can be used to generated electric power. Another variable that can be considered is the place, which is important taking into account electricity required, as mentioned in Reference [9]. In this work, wind farms were modeled near to thermal power plants and used their electric infrastructure. It was found the hours at day were both electricity demand and wind speed have their peaks, whit this information could benefit end-users as studied [12] that establish if consumers know the period of time were peak demand occurs could avoid higher electricity prices. This work presents a methodology to assess wind speed and calculate the power delivered to the grid, but is important to mention that if only wind speed is assessed and other variables such as roughness, orography, and climatology are not considered, wind cannot be considered as an unreliable source, as did Reference [15]. To avoid this problem, a proposal of cogeneration with thermal power plants has been made, in comparison with Reference [18] where a cost analysis was its main objective, we have analyzed wind data trajectory to reduce peak demand.

The proposal of connection to the power grid is based on the thermal power plant of Dos Bocas. The same connection is considered for the other thermal plants. In this case, the nominal power, that is, 350 MW, and the wind farm nominal power of 10.5 MW, is equivalent to 3% of the total substation capacity which indicates that the connection is possible. For Rio Bravo 800 MW, the connection will represent 1.3%, Altamira 1039 MW will be 1%, and Tuxpan 2263 MW. This connection represents 0.4% of the substation capacity.

Regarding the impact of new wind farms on power grids, Reference [35] established that power losses in distribution systems vary with numerous factors depending on the system configuration, such as level of losses through transmission and distribution lines, transformers, capacitors, and insulator, among others. In the case of this study, all the power losses were calculated in the original design. Appendix B shows a single line diagram where the transmission lines and the node modified with the connected wind farm can be seen.

4. Conclusions

The microscale model increased the accuracy of wind assessment to place wind farms. This is because when running the microscale model, several variables are taken into account such as roughness, orography, and climatology. In this study, four sites were assessed to model wind farms using MERRA-2 data, wind speed, wind direction, temperature, and atmospheric pressure, with records from 1980–2018.

The electricity that wind farms generate is delivered over transmission and distribution power lines. High-voltage transmission lines carry electricity over long distances to where consumers need it. More distance means more losses. The amount of power produced by each wind farm could contribute to the power produced by a thermal plant. In total for Northeast region they can deliver to the grid 2657.02 MW at peak demand. If there is 8000 MW demanded at this moment, the reduction will be at 5342.98 MW. At the Eastern region, the thermal power plant called Tuxpan is the biggest one in Mexico, so the amount of power in both wind farms and thermal plants is higher than Northeast. In this case, the total production is 3196.18 MW, and peak demand power delivered is 7200 MW if wind farms and thermal power plants work at the same time. The amount of power at this moment will be 4003.82 MW. There could be several techniques to contribute to decreased peak demand, energy storage technologies, and faster response times, but the most important is to nurture future generations.

The impact of the thermal power plants, once they have the contribution of wind energy, can be considered as follows: the efficiency for Rio Bravo and Altamira could be 31% and for Tuxpan and Dos Bocas 47%; the capacity factor, 36% and 56% respectively, however, the volatility and availability of wind in the entire year must be considered.

Mexico's main objective is to reach 35% of its electricity generated by clean sources in 2024. Employing this type of generation is important to contemplate both the trajectory of electricity demand and wind speed and to avoid a duck-curve that represents several renewable sources being contributed during the day but not at night, as solar or when wind does not has enough potential, e.g., Northeast curve has two peaks during the day, the first one peak is at 17:00 h and the second one is at 24:00 h, at this hour solar energy cannot be consider and if at this time does not flow the wind, the peak demand will increase and the prices will also do it.

Author Contributions: Q.H.-E. and A.-J.P.-M. have contributed to the theoretical approaches, simulation, experimental tests, and preparation of the article, and have contributed equally to this work. J.G., F.R.-M. and G.A. All authors have read and approved the final manuscript.

Funding: This research was funded by Ministry of Education Mexico, under the project Support to Research Groups, UV-CA-466.

Acknowledgments: Authors acknowledge to research group "Wind and solar energy" from Universidad del Istmo.

Conflicts of Interest: The authors declare no conflict of interest.

Appendix A

Table A1. Hourly mean wind speed in Rio Bravo Mexico.

Hour	Wind Speed [m/s]											
	Jan	Feb	Mar	Apr	May	Jun	Jul	Aug	Sep	Oct	Nov	Dec
0:00	7.1	7.7	8.4	8.5	8.6	8.6	8.8	8.2	7.1	7.3	7.2	6.9
1:00	7.2	7.8	8.3	8.3	8.3	8.3	8.4	7.8	7.0	7.3	7.3	7.0
2:00	7.1	7.6	8.0	7.9	7.8	7.8	7.8	7.3	6.6	7.0	7.1	6.9
3:00	6.9	7.3	7.7	7.6	7.4	7.4	7.4	6.9	6.3	6.7	6.9	6.8
4:00	6.8	7.1	7.4	7.3	7.2	7.1	7.1	6.5	5.9	6.4	6.8	6.7
5:00	6.6	7.0	7.2	7.2	7.1	6.8	6.8	6.2	5.6	6.2	6.6	6.6
6:00	6.6	7.0	7.1	7.1	7.0	6.6	6.6	5.9	5.3	6.0	6.5	6.5
7:00	6.5	6.9	7.1	7.1	6.9	6.4	6.4	5.6	5.1	5.9	6.5	6.5
8:00	6.5	6.9	7.1	7.0	6.9	6.3	6.2	5.3	4.9	5.8	6.5	6.4
9:00	6.5	6.9	7.1	7.0	6.8	6.1	6.0	5.1	4.8	5.7	6.5	6.5
10:00	6.6	6.9	7.1	7.0	6.8	6.0	5.8	5.0	4.8	5.6	6.5	6.5
11:00	6.6	6.9	7.1	7.0	6.8	5.9	5.7	4.9	4.7	5.6	6.4	6.5
12:00	6.6	6.9	7.1	6.9	6.6	5.7	5.5	4.7	4.7	5.6	6.4	6.4
13:00	6.5	6.7	6.8	7.1	7.3	6.6	6.4	5.2	4.5	5.3	6.3	6.4
14:00	6.1	6.7	7.6	7.8	7.8	6.9	7.1	5.9	5.1	5.6	6.0	5.9
15:00	6.6	7.4	8.0	7.9	7.8	6.9	7.1	5.9	5.3	6.2	6.8	6.5
16:00	7.0	7.4	7.9	7.7	7.6	6.7	7.1	5.9	5.4	6.2	7.0	7.0
17:00	7.0	7.2	7.7	7.5	7.4	6.7	7.1	5.9	5.4	6.1	6.9	6.9
18:00	6.8	7.0	7.5	7.3	7.3	6.8	7.2	6.1	5.6	6.1	6.7	6.7
19:00	6.7	6.9	7.4	7.3	7.3	7.1	7.5	6.5	5.8	6.1	6.5	6.5
20:00	6.6	6.8	7.4	7.4	7.5	7.4	7.9	6.9	6.1	6.3	6.5	6.4
21:00	6.5	6.9	7.5	7.6	7.8	7.8	8.3	7.3	6.4	6.5	6.5	6.4
22:00	6.5	7.1	7.8	7.9	8.1	8.2	8.6	7.8	6.7	6.8	6.6	6.3
23:00	6.7	7.4	8.1	8.3	8.5	8.5	8.8	8.1	6.9	7.0	6.8	6.5

Table A2. Hourly mean wind speed in Altamira Mexico.

Hour	Wind Speed [m/s]											
	Jan	Feb	Mar	Apr	May	Jun	Jul	Aug	Sep	Oct	Nov	Dec
0:00	6.7	7.1	8.0	8.0	7.2	6.3	5.7	5.4	5.7	6.1	6.3	6.5
1:00	6.7	7.1	7.7	7.6	7.0	6.2	5.7	5.4	5.6	6.1	6.3	6.5
2:00	6.5	6.8	7.2	7.1	6.7	6.1	5.5	5.2	5.5	5.9	6.2	6.4
3:00	6.1	6.3	6.6	6.5	6.3	5.8	5.3	5.0	5.1	5.6	5.9	6.1
4:00	5.7	5.8	6.0	5.8	5.7	5.5	5.1	4.7	4.7	5.3	5.5	5.7
5:00	5.3	5.3	5.4	5.2	5.1	5.1	4.8	4.3	4.3	4.9	5.2	5.3
6:00	5.0	4.9	4.9	4.7	4.6	4.7	4.4	3.9	4.0	4.6	4.9	5.0
7:00	4.8	4.6	4.6	4.3	4.2	4.2	4.0	3.5	3.7	4.3	4.7	4.9
8:00	4.7	4.4	4.3	4.1	3.9	3.8	3.6	3.2	3.5	4.2	4.6	4.8
9:00	4.7	4.4	4.1	3.9	3.7	3.5	3.3	2.9	3.5	4.1	4.5	4.8
10:00	4.7	4.4	4.1	3.9	3.6	3.3	3.1	2.8	3.5	4.1	4.5	4.8
11:00	4.8	4.4	4.1	3.9	3.5	3.2	3.0	2.7	3.5	4.1	4.6	4.8
12:00	4.8	4.5	4.1	3.9	3.5	3.1	2.9	2.7	3.6	4.2	4.6	4.8
13:00	4.9	4.5	4.2	3.9	3.6	3.1	2.8	2.6	3.5	4.2	4.7	4.8
14:00	4.8	4.6	4.3	4.1	3.8	3.2	2.9	2.7	3.5	4.1	4.6	4.8
15:00	4.8	4.7	4.5	4.2	3.9	3.3	3.0	2.7	3.5	4.1	4.6	4.7
16:00	4.8	4.7	4.6	4.4	4.1	3.4	3.1	2.8	3.6	4.1	4.7	4.7
17:00	4.8	4.7	4.8	4.6	4.4	3.7	3.4	2.9	3.7	4.1	4.6	4.7
18:00	4.8	4.8	5.0	5.1	4.9	4.1	3.7	3.3	3.9	4.3	4.6	4.6
19:00	4.8	5.1	5.5	5.6	5.3	4.5	4.2	3.7	4.3	4.5	4.7	4.7
20:00	5.1	5.5	6.1	6.3	5.9	5.0	4.7	4.2	4.7	4.8	5.0	4.9
21:00	5.5	6.1	6.9	7.1	6.4	5.5	5.2	4.7	5.1	5.2	5.4	5.3
22:00	6.1	6.7	7.6	7.7	6.9	5.9	5.5	5.1	5.4	5.6	5.8	5.8
23:00	6.4	7.1	8.0	8.1	7.1	6.2	5.7	5.3	5.6	5.9	6.1	6.1

Table A3. Hourly mean wind speed in Tuxpan Mexico.

Hour	Wind Speed [m/s]											
	Jan	Feb	Mar	Apr	May	Jun	Jul	Aug	Sep	Oct	Nov	Dec
0:00	6.0	6.5	7.0	7.2	7.2	6.6	6.0	5.6	5.5	5.8	5.9	5.9
1:00	5.8	6.2	6.5	6.7	6.8	6.4	5.9	5.5	5.5	5.7	5.8	5.7
2:00	5.4	5.7	5.9	6.1	6.2	5.9	5.6	5.2	5.3	5.5	5.6	5.5
3:00	5.1	5.2	5.4	5.5	5.7	5.5	5.3	4.9	4.9	5.2	5.3	5.2
4:00	4.9	4.9	5.1	5.1	5.2	5.0	4.9	4.4	4.5	4.8	5.0	5.0
5:00	4.9	4.8	4.9	4.9	4.8	4.5	4.5	3.9	4.2	4.6	4.9	5.0
6:00	4.9	4.7	4.8	4.7	4.5	4.1	4.1	3.6	4.0	4.5	4.8	4.9
7:00	4.9	4.7	4.6	4.5	4.3	3.8	3.8	3.4	3.9	4.5	4.8	4.9
8:00	4.9	4.7	4.6	4.4	4.2	3.6	3.6	3.2	3.9	4.5	4.8	4.9
9:00	4.9	4.7	4.6	4.4	4.1	3.4	3.4	3.1	4.0	4.6	4.8	4.9
10:00	4.9	4.7	4.6	4.4	4.0	3.3	3.2	3.0	4.0	4.6	4.8	4.9
11:00	4.9	4.7	4.5	4.3	3.9	3.2	3.1	3.0	4.0	4.7	4.8	4.9
12:00	4.9	4.7	4.5	4.2	3.7	3.1	3.0	3.0	4.0	4.6	4.8	4.9
13:00	4.9	4.5	4.3	4.1	3.7	3.1	3.0	2.9	3.9	4.5	4.7	4.8
14:00	4.7	4.5	4.4	4.1	3.7	3.2	3.1	2.9	3.8	4.3	4.6	4.7
15:00	4.7	4.6	4.4	4.1	3.7	3.2	3.1	2.8	3.7	4.3	4.6	4.7
16:00	4.8	4.7	4.6	4.3	3.8	3.4	3.3	2.9	3.8	4.3	4.6	4.7
17:00	4.8	4.8	4.8	4.7	4.2	3.8	3.6	3.2	3.9	4.4	4.7	4.7
18:00	4.9	5.0	5.2	5.1	4.8	4.4	4.1	3.6	4.1	4.5	4.8	4.8
19:00	5.1	5.4	5.7	5.7	5.4	5.1	4.6	4.1	4.4	4.7	4.9	4.9
20:00	5.3	5.7	6.2	6.3	6.0	5.7	5.2	4.6	4.7	4.9	5.1	5.1
21:00	5.6	6.0	6.7	6.8	6.5	6.2	5.6	5.0	5.1	5.2	5.3	5.3
22:00	5.8	6.4	7.0	7.2	6.9	6.5	5.9	5.4	5.4	5.5	5.6	5.5
23:00	5.9	6.5	7.2	7.3	7.2	6.7	6.0	5.5	5.5	5.6	5.7	5.7

Table A4. Hourly mean wind speed in Dos Bocas Mexico.

Hour	Wind Speed [m/s]											
	Jan	Feb	Mar	Apr	May	Jun	Jul	Aug	Sep	Oct	Nov	Dec
0:00	6.3	6.5	6.8	6.7	6.3	5.4	4.4	4.4	5.1	6.0	6.1	6.1
1:00	5.8	5.8	5.9	5.8	5.7	5.1	4.3	4.4	5.1	5.9	5.9	5.8
2:00	5.2	5.0	4.9	4.8	4.7	4.7	4.1	4.1	5.0	5.7	5.6	5.3
3:00	4.7	4.4	4.1	3.9	3.9	4.2	3.8	3.8	4.7	5.4	5.2	4.8
4:00	4.4	4.1	3.7	3.4	3.2	3.7	3.4	3.4	4.4	5.1	4.9	4.6
5:00	4.3	4.0	3.6	3.2	2.9	3.3	3.1	3.1	4.1	4.8	4.7	4.4
6:00	4.3	4.0	3.6	3.2	2.8	3.1	3.1	3.1	3.9	4.7	4.6	4.4
7:00	4.4	4.1	3.8	3.3	2.8	3.1	3.2	3.1	3.9	4.7	4.7	4.5
8:00	4.5	4.2	4.0	3.4	2.9	3.1	3.4	3.3	3.8	4.7	4.7	4.5
9:00	4.6	4.3	4.1	3.6	3.0	3.1	3.5	3.4	3.9	4.7	4.8	4.6
10:00	4.6	4.4	4.1	3.7	3.0	3.1	3.6	3.5	3.9	4.7	4.8	4.6
11:00	4.6	4.4	4.1	3.7	3.1	3.1	3.7	3.6	3.9	4.7	4.8	4.6
12:00	4.6	4.4	4.1	3.7	3.1	3.1	3.7	3.7	3.9	4.7	4.8	4.6
13:00	4.5	4.3	4.0	3.7	3.1	2.9	3.2	3.4	3.8	4.6	4.7	4.6
14:00	4.3	4.4	4.2	4.0	3.3	2.8	2.6	2.8	3.5	4.5	4.6	4.3
15:00	4.6	4.7	4.5	4.2	3.6	3.0	2.5	2.6	3.5	4.7	4.8	4.5
16:00	4.8	5.0	4.8	4.6	4.0	3.3	2.6	2.7	3.7	4.9	5.0	4.7
17:00	5.1	5.2	5.1	5.0	4.4	3.6	2.9	2.9	3.9	5.1	5.2	4.9
18:00	5.4	5.5	5.6	5.5	5.0	4.1	3.3	3.3	4.3	5.3	5.4	5.1
19:00	5.7	5.9	6.1	6.1	5.5	4.5	3.7	3.6	4.6	5.6	5.7	5.4
20:00	6.0	6.3	6.6	6.6	6.0	4.9	4.0	3.9	4.9	5.8	5.9	5.6
21:00	6.3	6.6	7.0	7.0	6.3	5.2	4.2	4.2	5.1	6.0	6.1	5.9
22:00	6.5	6.8	7.1	7.1	6.5	5.4	4.4	4.4	5.3	6.0	6.2	6.0
23:00	6.5	6.8	7.1	7.1	6.5	5.4	4.4	4.4	5.2	5.9	6.1	6.1

Appendix B

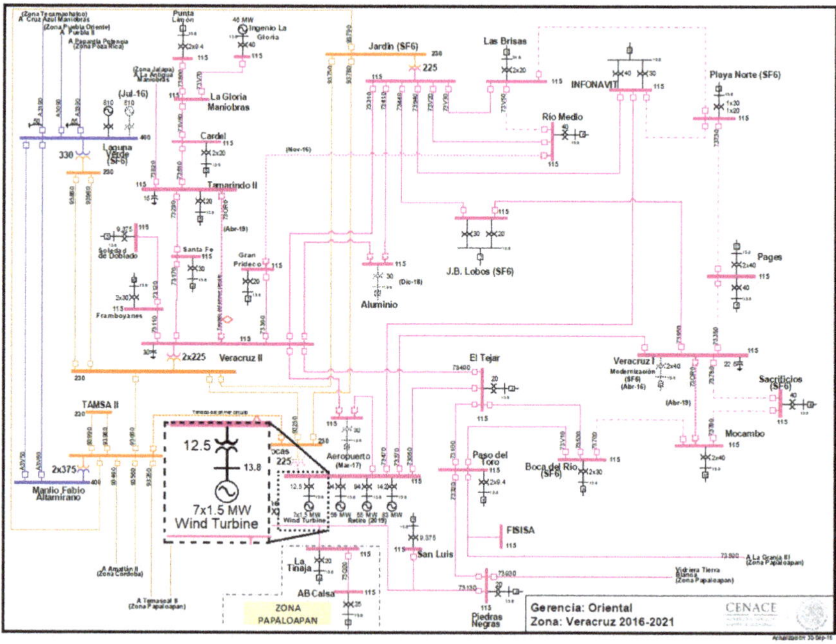

Figure A1. Single line diagram.

References

1. World Energy Statistics. 2018. Available online: https://webstore.iea.org/world-energy-statistics-2018 (accessed on 19 March 2019).
2. Electricity Information. 2018. Available online: https://webstore.iea.org/electricity-information-2018 (accessed on 19 March 2019).
3. Cui, W.; Li, L.; Lu, Z. Energy-efficient scheduling for sustainable manufacturing systems with renewable energy resources. *Nav. Res. Logist.* **2019**, *66*, 154–173. [CrossRef]
4. Adefarati, T.; Bansal, R.C. Reliability, economic and environmental analysis of a microgrid system in the presence of renewable energy resources. *Appl. Energy* **2019**, *236*, 1089–1114. [CrossRef]
5. Secretaría de Energía|Gobierno|gob.mx. Available online: https://www.gob.mx/sener (accessed on 20 March 2019).
6. DOF-Diario Oficial de la Federación. Available online: http://www.dof.gob.mx/nota_detalle.php?codigo=5421295&fecha=24/12/2015 (accessed on 21 March 2019).
7. SENER|Sistema de Información Energética. Available online: http://sie.energia.gob.mx/bdiController.do?action=temas (accessed on 21 March 2019).
8. Andersen, F.M.; Baldini, M.; Hansen, L.G.; Jensen, C.L. Households' hourly electricity consumption and peak demand in Denmark. *Appl. Energy* **2017**, *208*, 607–619. [CrossRef]
9. Fan, H.; MacGill, I.F.; Sproul, A.B. Statistical analysis of drivers of residential peak electricity demand. *Energy Build.* **2017**, *141*, 205–217. [CrossRef]
10. Lebotsa, M.E.; Sigauke, C.; Bere, A.; Fildes, R.; Boylan, J.E. Short term electricity demand forecasting using partially linear additive quantile regression with an application to the unit commitment problem. *Appl. Energy* **2018**, *222*, 104–118. [CrossRef]
11. Faruqui, A.; Hledik, R.; Newell, S.; Pfeifenberger, H. The Power of 5 Percent. *Electr. J.* **2007**, *20*, 68–77. [CrossRef]
12. Zaroni, H.; Maciel, L.B.; Carvalho, D.B.; de O. Pamplona, E. Monte Carlo Simulation approach for economic risk analysis of an emergency energy generation system. *Energy* **2019**, *172*, 498–508. [CrossRef]
13. Ikegami, T.; Urabe, C.T.; Saitou, T.; Ogimoto, K. Numerical definitions of wind power output fluctuations for power system operations. *Renew. Energy* **2018**, *115*, 6–15. [CrossRef]
14. Karimi, A.; Aminifar, F.; Fereidunian, A.; Lesani, H. Energy storage allocation in wind integrated distribution networks: An MILP-Based approach. *Renew. Energy* **2019**, *134*, 1042–1055. [CrossRef]
15. Waite, M.; Modi, V. Impact of deep wind power penetration on variability at load centers. *Appl. Energy* **2019**, *235*, 1048–1060. [CrossRef]
16. Saberi, K.; Pashaei-Didani, H.; Nourollahi, R.; Zare, K.; Nojavan, S. Optimal performance of CCHP based microgrid considering environmental issue in the presence of real time demand response. *Sustain. Cities Soc.* **2019**, *45*, 596–606. [CrossRef]
17. Watson, D.; Rebello, E.; Kii, N.; Fincker, T.; Rodgers, M. Demand and energy avoidance by a 2 MWh energy storage system in a 10 MW wind farm. *J. Energy Storage* **2018**, *20*, 371–379. [CrossRef]
18. Weicheng, S.; Wenxi, Z.; Dong, Z.; Yang, G. Wind Power Peak Regulation Pricing Model Under Wind and Fire Alternative Trading Mechanism–A Case Study of Wind Power Integration, Gansu Province (Apr 2018). In Proceedings of the 2018 China International Conference on Electricity Distribution (CICED), Tianjin, China, 17–19 September 2018; pp. 2736–2741.
19. Yan, X.; Lin, X.; Qin, L.; Han, S.; Gao, L.; Yang, Y.; Zeng, B. Control Strategy for Wind Power Integration Base on Energy Demand Respond and Distributed Energy Storage. *J. Eng.* **2017**. [CrossRef]
20. Long, H.; Xu, K.; Xu, R.; He, J. More Wind Power Integration with Adjusted Energy Carriers for Space Heating in Northern China. *Energies* **2012**, *5*, 3279–3294. [CrossRef]
21. Bexten, T.; Wirsum, M.; Roscher, B.; Schelenz, R.; Jacobs, G.; Weintraub, D.; Jeschke, P. Optimal Operation of a Gas Turbine Cogeneration Unit with Energy Storage for Wind Power System Integration. *J. Eng. Gas Turbines Power Trans.* **2019**, *141*. [CrossRef]
22. Wang, Y.-H.; Walter, R.K.; White, C.; Farr, H.; Ruttenberg, B.I. Assessment of surface wind datasets for estimating offshore wind energy along the Central California Coast. *Renew. Energy* **2019**, *133*, 343–353. [CrossRef]

23. Kim, D.; Ho, C.-H.; Park, D.-S.R.; Kim, J. Influence of vertical wind shear on wind- and rainfall areas of tropical cyclones making landfall over South Korea. *PLoS ONE* **2019**, *14*. [CrossRef]
24. Wang, M.; Ullrich, P.; Millstein, D. The future of wind energy in California: Future projections with the Variable-Resolution CESM. *Renew. Energy* **2018**, *127*, 242–257. [CrossRef]
25. PE|Electricidad. Available online: http://cuentame.inegi.org.mx/economia/parque/electricidad.html (accessed on 9 April 2019).
26. Caname. Available online: http://www.caname.org.mx/index.php/component/content/article?id=513 (accessed on 9 April 2019).
27. Tony, B.; Nick, J.; David, S.; Ervin, B. *Wind Energy Handbook*, 2nd ed.; Wiley: Hoboken, NJ, USA, 2011; ISBN 978-0-470-69975-1.
28. NECC (National Energy Control Center). CENACE. Available online: https://www.cenace.gob.mx/CENACE.aspx (accessed on 25 April 2019).
29. Demanda Regional (Regional Demand). Available online: https://www.cenace.gob.mx/Paginas/Publicas/Info/DemandaRegional.aspx (accessed on 25 April 2019).
30. Hussain, M.; Gao, Y. A review of demand response in an efficient smart grid environment. *Electr. J.* **2018**, *31*, 55–63. [CrossRef]
31. Nosratabadi, S.M.; Hooshmand, R.-A.; Gholipour, E. A comprehensive review on microgrid and virtual power plant concepts employed for distributed energy resources scheduling in power systems. *Renew. Sustain. Energy Rev.* **2017**, *67*, 341–363. [CrossRef]
32. Nikzad, M.; Mozafari, B. Reliability assessment of incentive- and priced-based demand response programs in restructured power systems. *Int. J. Electr. Power Energy Syst.* **2014**, *56*, 83–96. [CrossRef]
33. Wang, F.; Xu, H.; Xu, T.; Li, K.; Shafie-khah, M.; Catalão, J.P.S. The values of market-based demand response on improving power system reliability under extreme circumstances. *Appl. Energy* **2017**, *193*, 220–231. [CrossRef]
34. Sanz Rodrigo, J.; Chávez Arroyo, R.A.; Moriarty, P.; Churchfield, M.; Kosović, B.; Réthoré, P.-E.; Hansen, K.S.; Hahmann, A.; Mirocha, J.D.; Rife, D. Mesoscale to microscale wind farm flow modeling and evaluation. *Wiley Interdiscip. Rev. Energy Environ.* **2017**, *6*, e214. [CrossRef]
35. Augugliaro, A.; Dusonchet, L.; Favuzza, S.; Sanseverino, E. Voltage regulation and power losses minimization in automated distribution networks by an evolutionary multiobjective approach. *IEEE Trans. Power Syst.* **2004**, *19*, 1516–1527. [CrossRef]

© 2019 by the authors. Licensee MDPI, Basel, Switzerland. This article is an open access article distributed under the terms and conditions of the Creative Commons Attribution (CC BY) license (http://creativecommons.org/licenses/by/4.0/).

Article

Seasonal Wind Energy Characterization in the Gulf of Mexico

Alberto-Jesus Perea-Moreno [1], Gerardo Alcalá [2] and Quetzalcoatl Hernandez-Escobedo [3,*]

1 Departamento de Física Aplicada, Universidad de Córdoba, ceiA3, Campus de Rabanales, 14071 Córdoba, Spain; aperea@uco.es
2 Centro de Investigación en Recursos Energéticos y Sustentables, Universidad Veracruzana, Veracruz 96535, Mexico; galcala@uv.mx
3 Escuela Nacional de Estudios Superiores Juriquilla, UNAM, Queretaro 76230, Mexico
* Correspondence: qhernandez@unam.mx

Received: 21 November 2019; Accepted: 19 December 2019; Published: 23 December 2019

Abstract: In line with Mexico's interest in determining its wind resources, in this paper, 141 locations along the states of the Gulf of Mexico have been analyzed by calculating the main wind characteristics, such as the Weibull shape (*c*) and scale (*k*) parameters, and wind power density (WPD), by using re-analysis MERRA-2 (Modern-Era Retrospective Analysis for Research and Applications version 2) data with hourly records from 1980–2017 at a 50-m height. The analysis has been carried out using the R free software, whose its principal function is for statistical computing and graphics, to characterize the wind speed and determine its annual and seasonal (spring, summer, autumn, and winter) behavior for each state. As a result, the analysis determined two different wind seasons along the Gulf of Mexico;, it was found that in the states of Tamaulipas, Veracruz, and Tabasco wind season took place during autumn, winter, and spring, while for the states of Campeche and Yucatan, the only two states that shared its coast with the Caribbean Sea and the Gulf of Mexico, the wind season occurred only in winter and spring. In addition, it was found that by considering a seasonal analysis, more accurate information on wind characteristics could be generated; thus, by applying the Weibull distribution function, optimal zones for determining wind as a resource of energy can be established. Furthermore, a *k*-means algorithm was applied to the wind data, obtaining three clusters that can be seen by month; these results and using the Weibull parameter *c* allow for selecting the optimum wind turbine based on its power coefficient or efficiency.

Keywords: Weibull function; scale parameter; shape parameter; wind power density; seasons; Gulf of Mexico

1. Introduction

Global electricity demand grew 4% in 2018, almost twice than that for 2010, where renewables and nuclear power met most of the growth in demand [1]. According to the International Renewable Energy Association (IRENA) [2], 171 GW were added in 2018 worldwide after a strong growth in the last decade in renewable energy capacity. The total use of all renewables increased by 7.9%, where wind and solar energy contributed 84% of this total, and it is expected that in 2023, this increase for wind and solar energy will be 12.4% and 24%, respectively, in 2030, where the solar photovoltaic and wind power will be key energy sources since they are the energies with the highest growth, with the latter being one of the most profitable sources of energy in the world [3]. One of the most studied atmospheric parameters for decades is the direction of the wind [4]; nowadays this parameter is essential for the installation of a wind farm, where it is important that the wind turbines are not reducing their energy capacity due to poor design [5].

The role that Mexico will play on the renewable energy scenario is mainly outlined by its Energy Transition Law, which was recently reformed in 2015, establishing a minimum share of clean energy in the generation of electric power of 25% for 2018, 30% for 2021, and 35% for 2024 [6]. In 2017, there was a total electricity consumption of 260,051.895 GWh, out of which, renewables contributed around 77,907.2 GWh (30% of electric generation), where wind energy contributed 10,378 GWh (4% of electric generation), generated by 46 wind farms around Mexico [7].

Properly assessing wind resources for electricity generation implies knowledge of its behavior, which involves considering several variables that range from climate to the corrected wind turbine power curve (WTPC) selection [8]. Several studies in Mexico have located zones with suitable wind power, such as in the Baja California Peninsula, where it was found that the wind power density was above 400 W/m^2 [9]. Another study investigated the social impact caused by the expansion of large-scale wind energy projects on the Isthmus of Tehuantepec [10]. Furthermore, in the state of San Luis Potosi, a study of the installation of a hybrid PV-wind power generation system for social interest houses was validated [11]. Also, a statistical methodology based on support vector regression for wind-speed forecasting located at La Ventosa, Oaxaca, Mexico, showed that forecasts made with their own method are more accurate for medium (5–23 h ahead) short-term wind speed forecasting (WSF) and wind-power forecasting (WPF) than those made with persistent and autoregressive models [12] and a wind power map of all of Mexico [13].

According to Mazzeo et al. [14], numerous studies have been done to classify patterns of weather variables, where some are related to cluster analysis (CA) [15,16] and other to the similarity of time-series wind vectors [17]. CA has been successfully applied to the regionalization of wind in complex terrain, such as those of the north of the Iberian Peninsula [18]. Furthermore, in the long-term, it will be useful to detect wind patterns through the analysis of their time series with fast Fourier transform (FFT) [19] or wavelet methods [20]. The use of the latter technique has proved useful for missing wind data arrangement for a long time series of wind data [21].

Therefore, the crucial step in wind assessment is to determine the wind resources, which depends on accurate wind-speed modelling [22]. One of these models is the probability density function (PDF) because it provides important wind speed distribution parameters and allows one to determine the Weibull parameters [23]. It is known that if wind speed follows the Weibull distribution with scale parameter (c), which has the same units as wind speed and the dimensionless shape parameter (k), the load and power density also follow the same distribution with shape parameters $k/2$ and $k/3$, respectively, and scale parameters $c/2$ and $c/3$, respectively [24]. To calculate the Weibull parameters, some processes have been developed, e.g., Saleh et al. [24] reviewed six kinds of numerical methods commonly used for estimating the Weibull parameters: the moment, empirical, graphical, maximum likelihood, modified maximum likelihood, and energy pattern factor methods. From the review, they found that if the wind speed distribution matched well with the Weibull function, the six methods were applicable; but if not, the maximum likelihood method performed best, followed by the modified maximum likelihood and moment methods. Furthermore, Aukitino et al. [25] assessed the wind speed and found out that moment method was the best fitting, Akdağ and Dinler [26] proposed a new method based on wind power density and mean wind speed called the power density (PD) method, Baseer et al. [27] estimated the Weibull parameters using a least-squares regression method (LSRM) and the Wind Atlas Analysis and Application Program (WAsP) algorithm, and Ozay and Celiktas [28] did an statistical analysis in Turkey using WAsP.

Wind power density (WPD) is one of the most important factors to consider when wind is assessed to generate power and depends on an accurate determination of the Weibull parameters. The literature shows the relationship between the WPD and the Weibull function. Katinas et al. [29] performed a study in Lithuania where the WPD was calculated after determining the Weibull parameters. In Spain, an evaluation of the WPD was done using the moment method for calculating the Weibull parameters [30]. Faghani et al. [31] extrapolated data at high altitudes, calculated the

Weibull parameters, and found out that variation of power density with time was significant; therefore, they divided the year in two periods, period I (spring and summer) and period II (autumn and winter).

The ideas of Faghani et al. [31] are considered for this study and are extended for seasonal analysis (spring, summer, autumn, and winter) to determine its Weibull parameters and the characteristics of wind speed. To achieve this, a statistical analysis was conducted, and with this information, wind can be utilized effectively as a resource for electric generation.

The main objective of this study was to identify seasonal wind characteristics to assess them and determine their potential using different types of wind turbines based on their power curves and power coefficients. We considered a very important step in the process of wind turbine selection, which is the assessment or characterization of the wind speed. In this study, we also include a proposal to relate the wind turbine efficiency through its power coefficient and the conditions of the wind at each specific site.

2. Materials and Methods

2.1. Sites and Data

Mexico has five states along the Gulf of Mexico: Tamaulipas, Veracruz, Tabasco, Campeche, and Yucatan, as seen in Figure 1.

In order to carry out this study, data for 141 different sites along the Mexican Gulf were collected from the Modern-Era Retrospective Analysis for Research and Applications version 2 (MERRA-2) from the National Aeronautics and Space Administration (NASA); these are re-analysis data that are long-term, model-based analyses of multiple datasets using a fixed assimilation system [32]. MERRA-2 has presented data every hour from 1980 until 2017. According to Shang [33] the network measures derived from the empirical observations are often poor estimators of the true structure of system as it is impossible to observe all components and all interactions in many real-world complex systems. This problem occurs when there is missing data; in this study, MERRA-2 has no missing data; therefore, this problem is avoided and can be considered to be unbiased data.

In Figure 2, we give the geographic positions where MERRA-2 data were taken from.

Figure 1. Mexican states along the Gulf of Mexico.

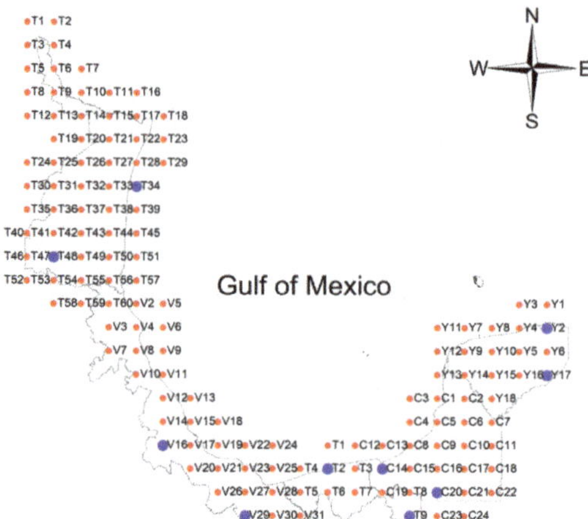

Figure 2. Modern-Era Retrospective Analysis for Research and Applications version 2 (MERRA-2) data locations.

2.2. The Weibull Distribution and Seasonal Wind

There are some statistical distribution functions for analyzing wind data, such as the lognormal, normal, Rayleigh, and Weibull probability distributions [34,35]. The Weibull function is the most-used function to assess wind energy potential because shows variables as shape and scale parameters [36], where these parameters are obtained using estimation methods, such as the maximum likelihood method, and the goodness of the resulting fits are evaluated using several indicators, e.g., the coefficient of determination (R^2) [24,37–39]. The R^2 is a statistical measure that represents the proportion of the variance for a dependent variable that is explained by an independent variable; R^2 is generally interpreted as the percentage of a value's movements that can be explained by movements of another variable [40].

2.2.1. Wind Model

The Weibull distribution and cumulative distribution functions are expressed in Equations (1) and (2), respectively [41]:

$$f(v) = \frac{k}{c}\left(\frac{v}{c}\right)^{k-1} exp\left[-\left(\frac{v}{c}\right)^{k}\right](k > 0, v > 0, c > 1), \quad (1)$$

$$F(v) = 1 - exp\left[\left(\frac{v}{c}\right)^{k}\right], \quad (2)$$

where k is the shape parameter (dimensionless), which is considered a Weibull form parameter because it specifies the shape of the distribution taking place within values between 1 and 3. A small value for k signifies very variable winds, while constant winds are characterized by a larger k [42]; when the shape parameter is 2, it is considered to represent Rayleigh distribution. The scale parameter c has the same units as wind speed (m/s) and is proportional to the mean wind speed (v_m), where v (m/s) is the wind speed registered in the site.

If the mean wind speed in Equation (3), and the standard deviation (σ) are known, k and c can be calculated using Equations (4) and (5) as follows [43]:

$$v_m = \frac{1}{n}\left[\sum_{i=1}^{n} v_i\right], \quad (3)$$

$$k = \left(\frac{\sigma}{v_m}\right)^{-1.086}, \tag{4}$$

$$c = \frac{v_m}{\Gamma\left(1 + \frac{1}{k}\right)}. \tag{5}$$

The gamma function, Γ, can be calculated using Equation (6):

$$\Gamma(r) = \int_0^\infty x^{r-1} e^{-x} dx. \tag{6}$$

2.2.2. Wind Power Analysis

The observed wind power density (WPD_O) can be obtained using Equation (7):

$$WPD_O = \frac{\sum_{i=1}^{n} 0.5 \rho v_i^3}{n}, \tag{7}$$

where ρ is the air density and is calculated using the ideal gas law (see Equation (8)), in which T is the absolute temperature (K), p is the absolute pressure (Pa), and R is the specific gas constant (J/kg·K): As an approximation, $R = 0.286$ is used for dry air.

$$\rho = \frac{p}{RT} \tag{8}$$

2.3. Wind Variables Analysis

Calculations for wind variables were performed by implementing a self-written code on the R free software environment and language [44]. This environment, besides being free, provides powerful tools for statistical computing and graphical display via eight packages, and when required, they can be extended with additional packages available through the comprehensive R archive network (CRAN) family available on the Internet. In fact, in the former study, six base packages were used (tools, stats, graphics, grDevices, utils, and base) [44], which were complemented with the extra packages (lubridate [45], RColorBrewer [46], ggplot2 [47], and gridExtra [48]).

2.3.1. Data Processing

An iterative process was run for all 141 studied sites, which were distributed in a grid-shaped arrangement along the coast of the Gulf of Mexico (see Figure 2) using the previously mentioned MERRA-2 files as input data. These files contained complete time series data (no missing values) of the surface pressure (kPa), air temperature (°C) at 2 m and 10 m above ground level, wind speed (m/s) at 50 m above ground level, and wind direction (°) at 60 m above ground level; all these variables have hourly records from 1980 to 2016.

An annual and a seasonal analysis was performed for every site, where for the sake of simplicity, spring was considered to include the months of March, April, and May; summer included June, July, and August; autumn included September, October, and November; and Winter included December, January, and February.

The structure of the program consisted of an initial pre-processing of the MERRA-2 data, which generated an output file that was used for obtaining the geolocated values of the wind variables, as well as customized plots via two scripts named Weibull Analysis and Directional Analysis, as can be observed on Figure 3, and is explained as follows.

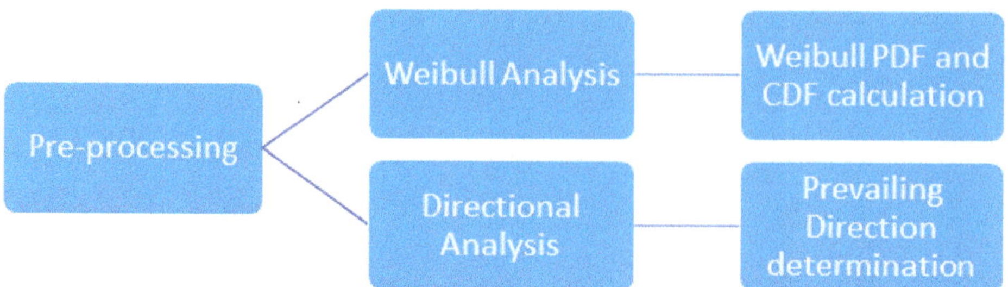

Figure 3. Program structure. CDF: cumulative density function, PDF: probability distribution function.

2.3.2. Pre-Processing

In the pre-processing step, a time adjustment was performed on the MERRA-2 data in order to assign the corresponding UTM zone (for Mexico −6 h) and make summertime corrections. Data was classified for years, months, days, and hours for further manipulation, and the air density and observed WPD were obtained via Equations (9) and (10), respectively.

Once the pre-processing was finished, an output was generated for feeding the following two independent and complementary scripts.

2.3.3. Weibull Analysis

The Weibull probability distribution function (PDF) and cumulative distribution function (CDF) were obtained and plotted according to Equations (1) and (2), respectively, for the annual and seasonal series.

Tables were generated with Weibull shapes and scale factors (k, c), mean wind speed (v_m), its corresponding standard deviation (σ), Weibull most-probable wind speed (v_{mp}), mean air density (ρ_m), and the observed wind power density (WPD_O).

2.3.4. Directional Analysis

In order to complement the Weibull distribution, a directional analysis was performed. For this, data was grouped according to the wind direction into 12 bins, corresponding to sectors as 30° segments. The cumulative WPD_O was obtained for every sector by considering the one with the highest value as the prevailing direction, which was obtained for complete years, as well as for every season.

2.4. *Wind Clustering*

The wind performance was grouped into several clusters using a *k*-means clustering model to identify its monthly behavior, which was analyzed to diagnose the time where the wind speed could be used to generated power. The *k*-means algorithm is widely used in data mining for the partitioning of n measured quantities into k clusters [49]; according to Sugar and James [50], the classification of observations into groups requires computing the distance between the observations [51–54]. The *k*-means algorithm is one of the simplest unsupervised machine learning algorithms, where unsupervised algorithms make inferences from datasets using only an input vector without referring to known, or labelled, outcomes [53]. We can define a cluster as a data set with similar characteristics. The *k*-means algorithm identifies k centroids and then allocates every data point to the nearest cluster [53]. This algorithm has been used in a study done by Wang et al. [55], where the *k*-means clustering algorithm was used to find the largest historical samples that had the greatest influence on forecasting accuracy to improve the efficiency of the proposed model.

A method for clustering was proposed by Deng et al. [52], which used a Weibull distribution to establish that an unclustered dataset P can be represented using Equation (9):

$$P = \left\{p_i,\ i = 1,2,3,\ldots,N^i\right\}, \tag{9}$$

where p_i denotes the ith element and N^i is the number of observations. The set of clusters C can be given as:

$$C = \left\{C_j \middle| C_j \subset P,\ j = 1,2,3,\ldots,N^j\right\}, \tag{10}$$

and the set of observations within a cluster, C_j, is represented by:

$$C_j = \left\{C_{jk} \middle| C_{jk} \subset P,\ j = 1,2,3,\ldots,N^{jk}\right\}. \tag{11}$$

C_{jk} and N_{jk} are the kth observation with the jth cluster and the number of observations, respectively, in cluster set C_j. The set of centroids associated with the clusters W is given as:

$$W = \left\{W_j \middle| W_j \subset C_j,\ j = 1,2,3,\ldots,N^j\right\}, \tag{12}$$

where W_j is the jth centroid.

For wind clustering, each observation P is assigned to a cluster C_j and its centroid will be represented by its mean wind speed.

2.5. Wind Turbine Selection

The energy available for conversion mainly depends on the wind speed and the swept area of the wind turbine. Using Newton's Law $F = ma$:

$$E = mas, \tag{13}$$

where E is the kinetic energy, m is the mass, a is the constant acceleration, and s is the distance. From kinematics, the acceleration can be expressed using Equation (14):

$$a = \frac{\left(v^2 - u^2\right)}{2s}. \tag{14}$$

Because the initial velocity of the object is 0, then $u = 0$, such that:

$$a = \frac{v^2}{2s}. \tag{15}$$

Substituting Equation (15) in Equation (13) gives:

$$E = \frac{1}{2}mv^2. \tag{16}$$

The rate of change of energy of the power in the wind is given by:

$$P = \frac{dE}{dt} = \frac{1}{2}v^2\frac{dm}{dt}. \tag{17}$$

The mass flow rate is expressed as:

$$\frac{dm}{dt} = \rho A\frac{dx}{dt}, \tag{18}$$

and the distance's rate of change is:

$$\frac{dx}{dt} = v. \tag{19}$$

Substituting Equation (18) into Equation (19) gives:

$$\frac{dm}{dt} = \rho A v. \tag{20}$$

Hence, from Equation (17), the wind power output generated by a wind turbine can be expressed using Equation (21):

$$P_{WT} = \frac{1}{2}\rho A U^3, \tag{21}$$

where P_{WT} is the rated power, ρ is the air density, A is the rotor area, and U is the wind speed approaching the turbine. According to Grillo et al. [56], the power coefficient is a function of the tip speed ratio, known as lambda (λ), and the blade pitch angle (β) (°). λ can be calculated Equation (22):

$$\lambda = \frac{\omega R}{U}, \tag{22}$$

where ω is the rotational speed of the rotor (rad/s) and R is the rotor radius (m).

To characterize the wind speed in this study, a proposal for wind turbine selection is given, where this proposal considers the air density, rotor area, wind turbine power curve, and power coefficient (C_p).

Song et al. [57] defined the power coefficient C_p as the ratio of the power extracted from the wind turbine to the available power. C_p can be calculated as follows:

$$C_p = \frac{P_n}{\frac{1}{2}\rho A u_j^3}. \tag{23}$$

The C_p of a wind turbine is a measurement of how efficiently the wind turbine converts the energy in the wind into electricity.

Wind speed is one of the most important parameters in determining the electric power, and the general equation is related to the density of air, wind speed, and swept area, as in Equation (21), and represents the total energy obtained from the wind resource; however, in terms of generating electricity, only a certain proportion of energy can be converted and is expressed by Equation (24), as follows:

$$P_e = \eta_e \eta_m C_p P_{wT}, \tag{24}$$

where P_e is the amount of electric power generated, η_e is the electrical conversion efficiency of the wind turbine, and η_m is the mechanical efficiency [58].

There is an optimization proposal that uses the wind turbine efficiency as a fundamental variable to determine the power output generated by a wind power farm [59]. In this proposal, as in this study to determine it, a Weibull distribution was used. The parameter c from the Weibull distribution was used to select a wind turbine according to its C_p; in this case, the maximum efficiency of a wind turbine was compared to the parameter c calculated from the wind speed site studied.

3. Results and Discussion

The annual Weibull parameters c and k, WPD_F, WPD_O, ρ, v_m, v_{mp}, and v_{maxE} were obtained for the 141 MERRA-2 data locations. Table 1 only presents the places with the highest and lowest mean wind speeds for each state (colored blue in Figure 2); these values were used as references for the seasonal analysis.

Table 1. MERRA-2 analysis.

State	ID	c	k	WPD_O	ρ	v_m	v_{mp}
		(m/s)	(–)	(W/m²)	(kg/m³)	(m/s)	(m/s)
Tamaulipas	Tam34	8.26	2.71	357.09	1.19	7.34	6.97
Tamaulipas	Tam48	4.18	2.27	48.32	1.10	3.71	3.24
Veracruz	Ver16	3.86	2.23	37.82	1.06	3.42	2.96
Veracruz	Ver29	6.46	2.40	182.18	1.17	5.72	5.16
Tabasco	Tab2	6.90	2.84	200.40	1.18	6.15	5.93
Tabasco	Tab9	4.02	2.76	40.02	1.17	3.58	3.42
Campeche	Cam14	6.51	3.08	161.68	1.18	5.82	5.73
Campeche	Cam20	2.90	3.04	14.14	1.17	2.59	2.54
Yucatan	Yuc2	5.10	3.07	79.38	1.19	4.56	4.48
Yucatan	Yuc17	2.99	3.10	15.76	1.18	2.67	2.63

Tamaulipas was the state with more zones with high values of wind than the other states, where its maximum v_m = 7.34 m/s, scale and shape parameters were between 4.17–8.26 m/s and 1.97–2.91, respectively. In this case, by comparing v_m, c, and k, it can be established that in the time domain, most of the wind resource had values higher than its mean wind speed. In Veracruz, the scale and shape parameters showed values between 3.86–6.88 m/s and 1.72–2.41, respectively. Tabasco presented more wind speed and wind resource than Veracruz with Weibull parameters ranging between 4.02–7.48 m/s for c, and 2.51–3.07 for k. Campeche and Yucatan showed a higher k's than the others states, with 2.71 and 2.95, respectively, which can be interpreted as a long period of time with winds higher than their means; c was 7.47 m/s and 7.88 m/s for Campeche and Yucatan, respectively.

According to Katinas et al. [29], the Weibull parameter c has the same behavior as the WPD, where in all cases, when the scale factor grows, the WPD grows as well. It is important to mention that the lowest shape parameter corresponds to the lowest wind speed, which means that the frequency of data below the mean was greater than that above it.

3.1. Seasonal Weibull Parameters and WPD

Seasonal wind speed characterization was carried out by dividing the data according to the four annual seasons: spring (March, April, and May); summer (June, July, and August); autumn (September, October and November); and winter (December, January and February). For the sake of analysis, two sites for each state were chosen by considering its highest and lowest mean wind speed (see Figures 4–8).

In Figure 4, it can be observed that the points Tam34 and Tam48 had the highest and the lowest mean wind speed—8.26 m/s and 4.18 m/s, respectively—in the state of Tamaulipas. At the annual level, Tam34 had k = 2.71, which represents a very good value due its higher magnitudes of wind speed; at the seasonal level, the Weibull distribution shows that spring and winter are the periods of time during an average year where there was more available wind resource.

In the state of Veracruz, 31 locations were studied, with the highest scale parameter being found for Ver29, where this was located is in the Isthmus of Tehuantepec, one of the windiest zones in the world. During an average year, its highest mean wind speed was 5.72 m/s, its c was 6.46 m/s, and k was 2.40, and had a windy season during spring, autumn, and winter (see Figure 5).

In Tabasco (Figure 6), Tab2 had c = 6.90 m/s, k = 2.84, and a mean wind speed of 6.15 m/s; these values represent a good wind resource because analyzing Weibull parameters showed that c was higher than its mean, and k showed that most of the time, the wind speed data was above its mean. In contrast, the seasonal Weibull of Tab9, which was the point with the lowest values in this state (c = 4.02 m/s, k = 2.76, v_m = 3.58 m/s), allowed to determine a period of wind during an average year that had constant magnitudes of wind values, namely winter, spring, and summer, as shown.

The analysis of Campeche (Figure 7) showed its windiest location was Cam14, which had a mean wind speed of 5.82 m/s, and c and k values equal to 6.51 m/s and 3.08, respectively. The lowest valued location, Cam20, had c = 2.90 m/s and v_m = 2.64 m/s, where both values are similar with a difference of

0.26 m/s, which can be considered as a location that did not have variations during an average year. Its k = 3.04 could be considered a very good frequency of wind speed; however, due its low wind speed, it did not represent an impactful wind resource, which can be seen in the seasonal analysis because its variation was minimal.

In Yucatan, as seen in Figure 8, its seasonal wind variation was divided in two periods, highest one between winter and spring and the lowest one between summer and autumn. The location with the highest wind speed was Yuc2 (v_m = 4.56 m/s, c = 5.10 m/s, and k = 3.07), and the lowest location, Yuc17, was inside a rainforest, which reduced the magnitude of the wind speed (v_m = 2.67 m/s, c = 2.99, and k = 3.10).

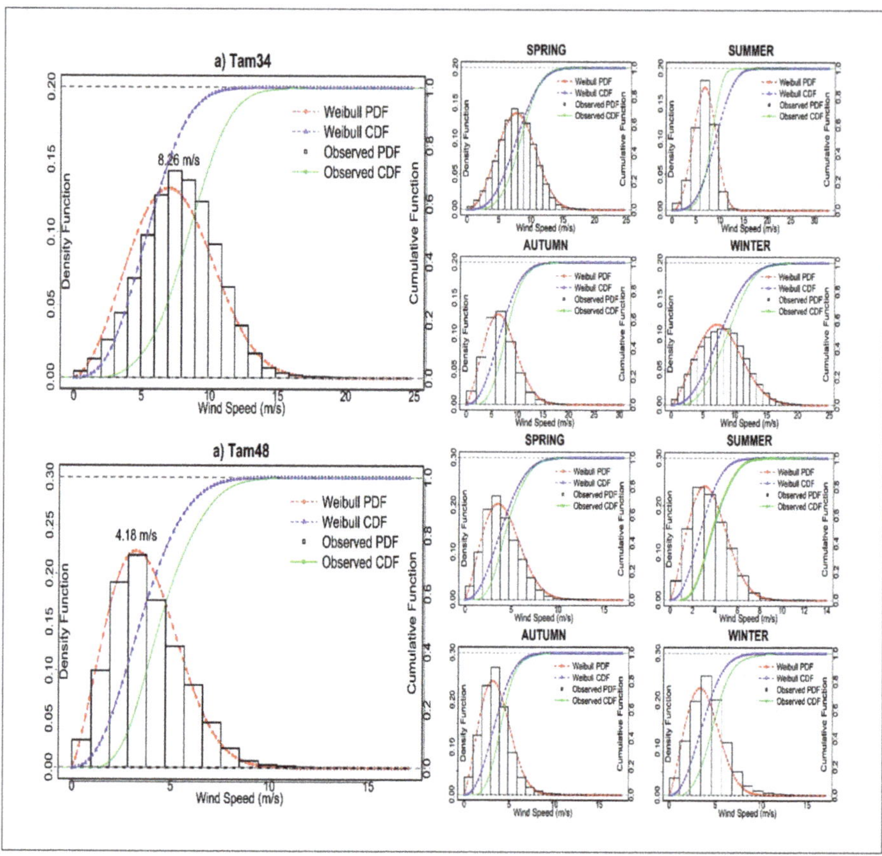

Figure 4. Annual and seasonal Weibull distributions in the state of Tamaulipas.

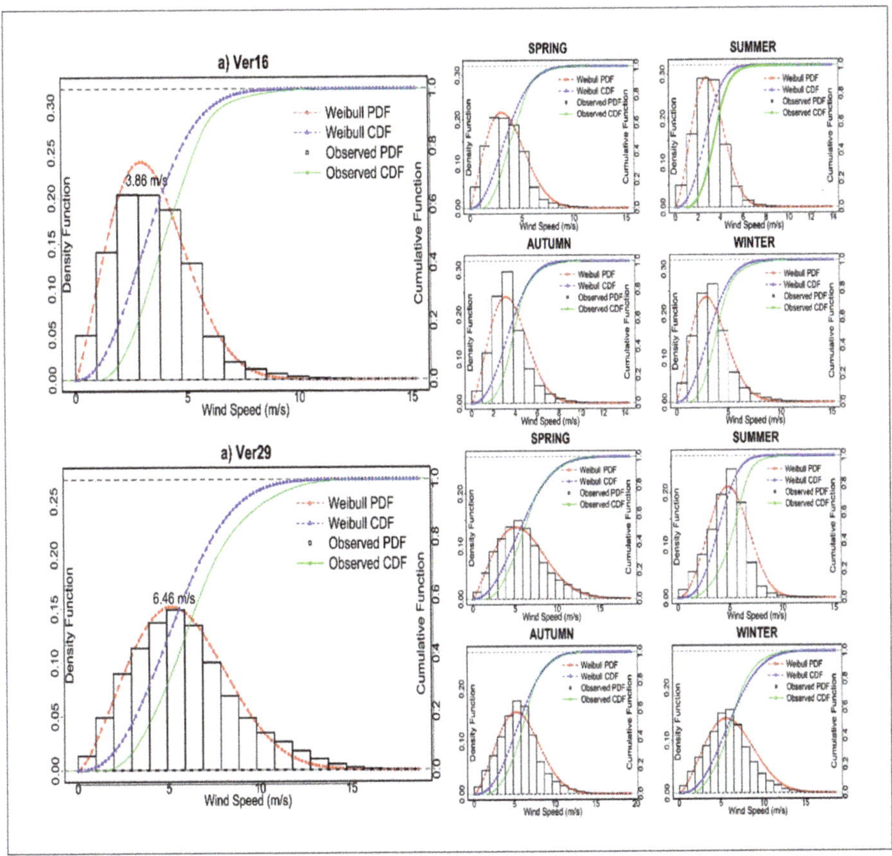

Figure 5. Annual and seasonal Weibull distributions for the state of Veracruz.

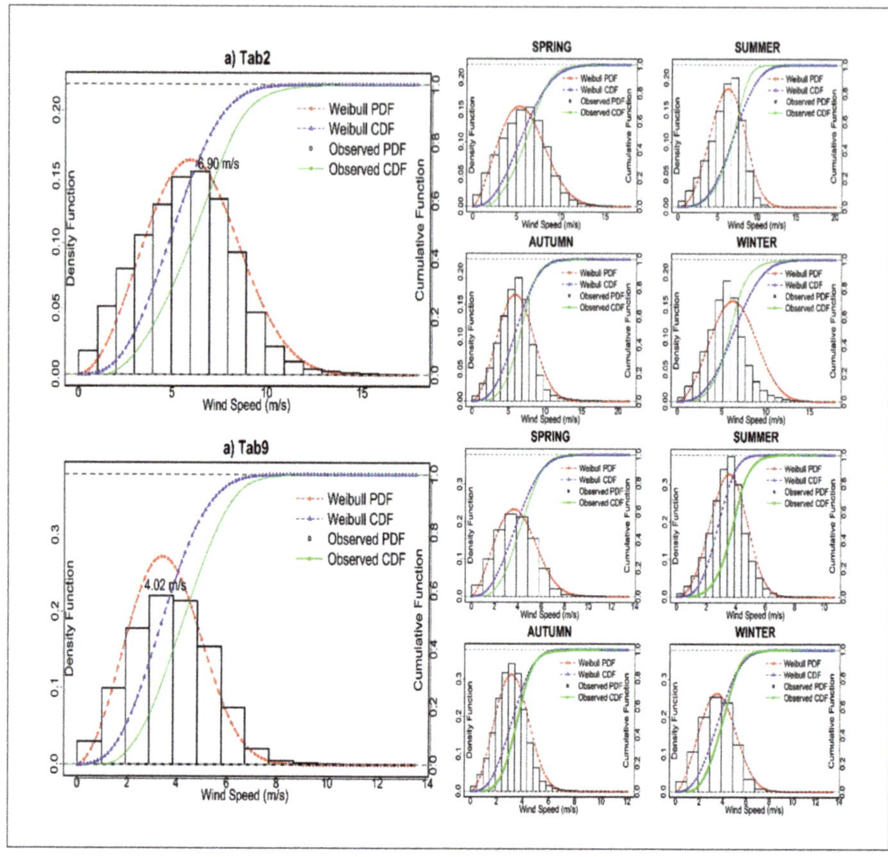

Figure 6. Annual and seasonal Weibull distributions for the state of Tabasco.

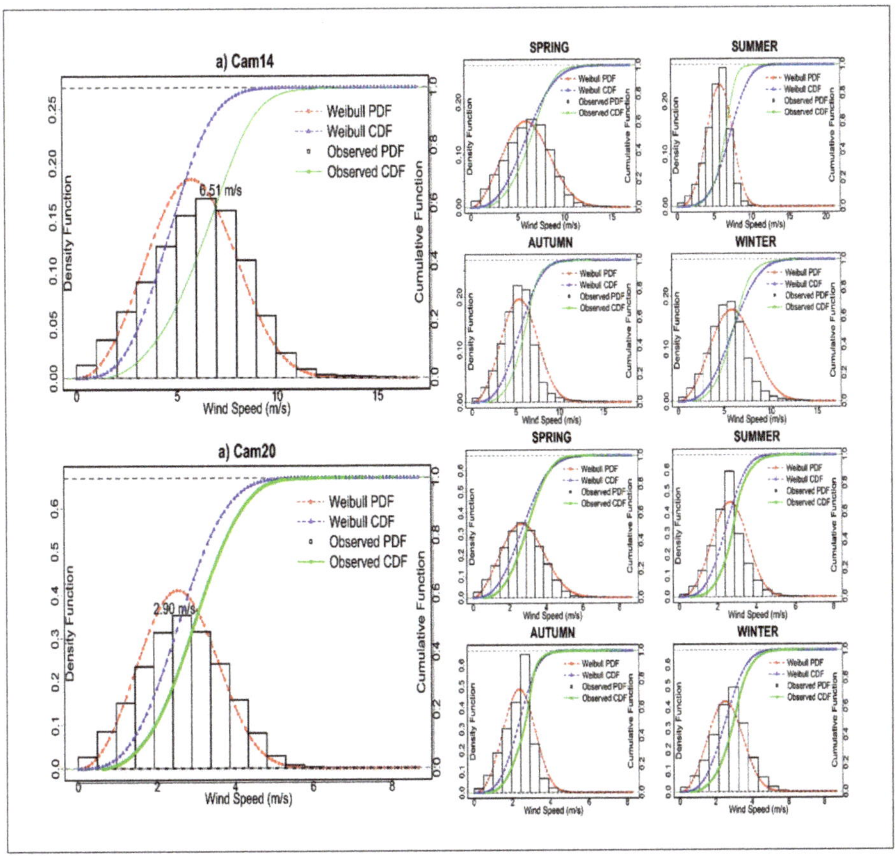

Figure 7. Annual and seasonal Weibull distributions for the state of Campeche.

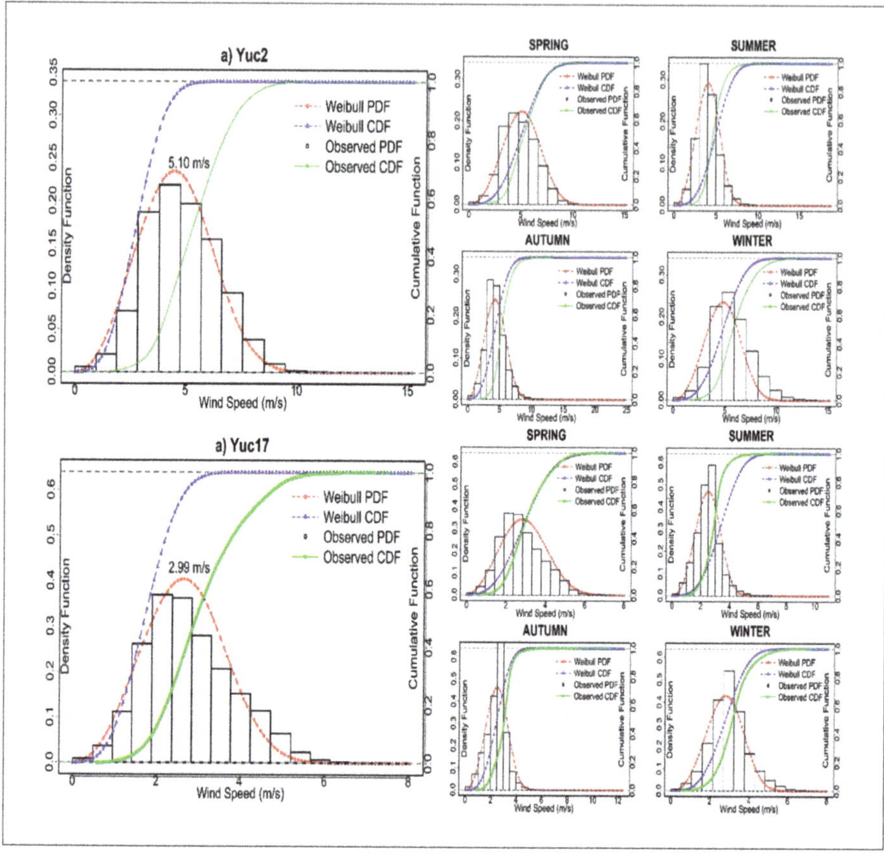

Figure 8. Annual and seasonal Weibull distributions for the state of Yucatan.

These results are in agreement with the study developed by Herrero-Novoa [30], where it was established that dividing an average year into two periods (spring–summer and fall–winter) provides more accuracy regarding wind characterization; in a similar fashion, for the current study, an average year was divided into four periods of time (seasons) where even more accurate wind periods were established. This allowed us to determine the wind characteristics, specifically its WPD, in the most relevant period. By observing Figure 8 and considering the values of the parameters obtained in Table 1, we established that in Yuc2, the parameter c = 5.10 m/s was higher than its average of 4.56 m/s, and its k = 3.07; therefore, at this site, the wind speed was over 4.56 m/s most of the time. Furthermore, according to Vazquez et al. [42], this value of k means that the site had constant winds.

In addition to the Weibull analysis, in Figures 9 and 10, the prevailing wind directions and the magnitude of its WPD_O can be observed for the annual and seasonal behavior, respectively.

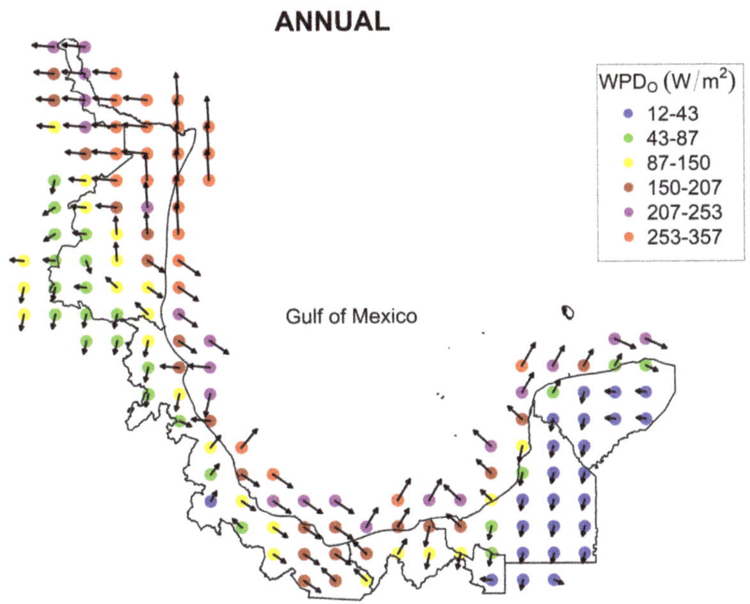

Figure 9. Annual prevailing wind direction and its WPD_O along the states of the Gulf of Mexico.

The highest WPD_O points were along Tamaulipas and Veracruz, although Tamaulipas had more locations than Veracruz, while Veracruz has the highest resource next to the sea, Tamaulipas had it at the border with the United States of America.

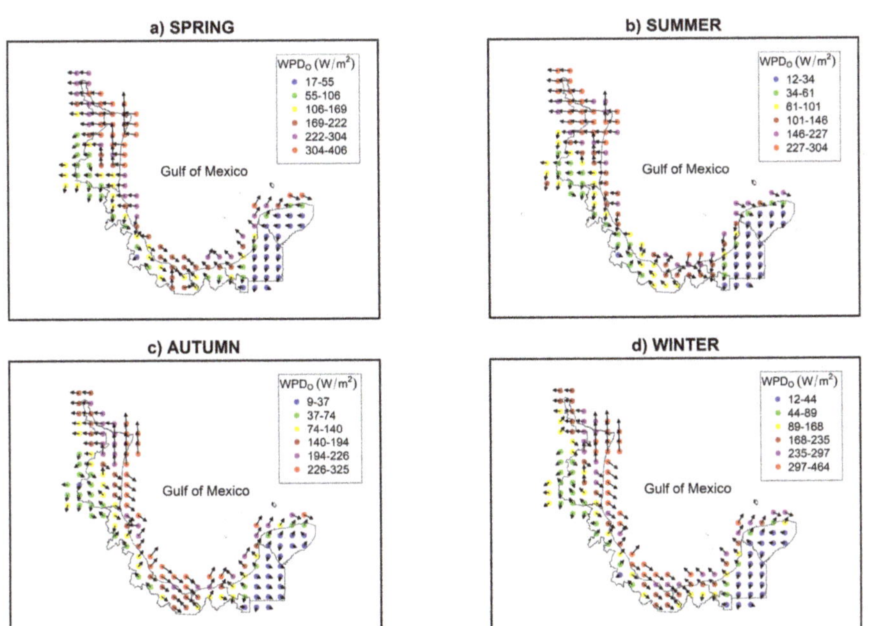

Figure 10. Seasonal prevailing wind direction and its WPD_O along the states of the Gulf of Mexico.

As can be observed in Figure 10, the seasons with greatest WPD_O were autumn and winter, and the states with greatest resource were Tamaulipas and Veracruz. Regarding wind orientation, this figure shows that Tamaulipas had two periods with different prevailing directions, with the first one during spring and summer and the second one in autumn and winter. The WPD in southern Veracruz presented the same direction most of the time (southeast), meanwhile Tabasco, which is the southernmost state along the Gulf of Mexico, had different prevailing orientations throughout an average year. It can also be seen that Campeche had a prevailing direction in winter and spring, and for Yucatan, autumn, winter, and spring had northeast as the predominant direction, while it was southeast in summer.

3.2. Wind Speed and Wind Power Clustering

Using *k*-means clustering, the wind speed could be clustered into months, and in these clusters, the power output could be calculated. Figure 11 shows the clustering done for wind speed from 1980–2017.

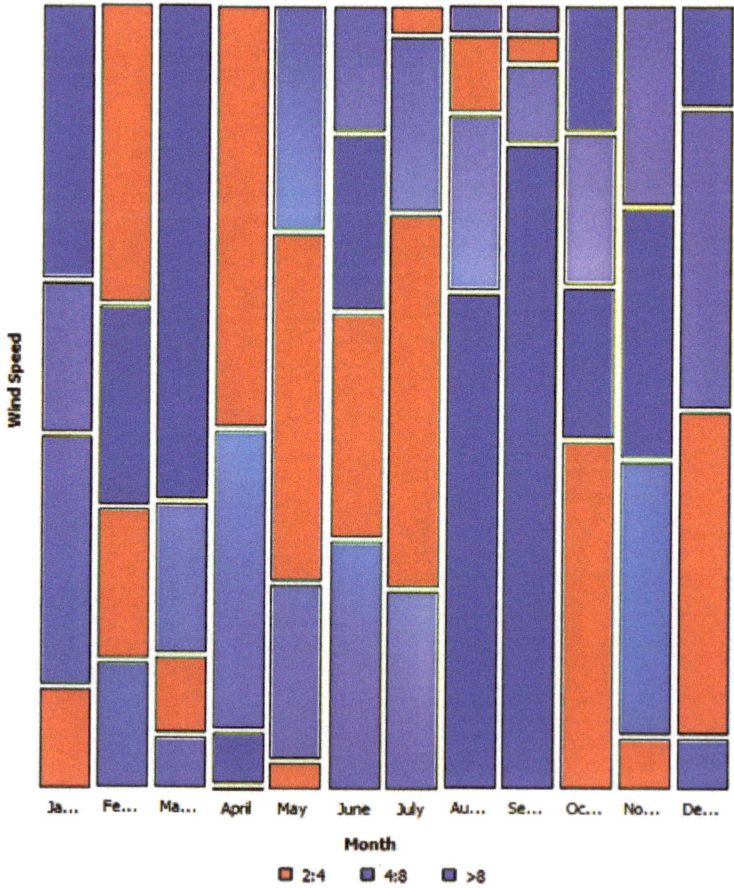

Figure 11. Wind speed clustering in the Mexican states of the Gulf of Mexico.

Wind speed was divided into three clusters—C_1, C_2, and C_3—as shown in Figure 11 where the blocks represent the amount of wind speed each month.

Taking these results and knowing the windy seasons throughout an average year, the wind power output could be calculated. A critical point in this stage was to select the wind turbine, where in Table 2, we present the C_p of different wind turbines as a first step to know the efficiency of each one of them. The wind turbines were selected according to parameter c of the points analyzed.

Table 2. Wind turbine C_p.

Wind Turbine	C_p	Wind Speed (m/s) @ Cp
Acciona AW70/1500 Class III	0.4188	6.5
Acciona AW70/1500 Class II	0.4457	7.5
Ecotecnia 48	0.4142	8.5
Clipper Liberty C93	0.4215	9.5
Nordex S70 1500 kW	0.4410	10
Vestas V90 1.8 MW	0.4381	8

Six wind turbines were found such that their C_p fit with the parameter c calculated from the wind speed data. Table 3 presents the wind turbine selected for each point along the Gulf of Mexico.

Table 3. Wind turbine selection.

ID Point	c (m/s)	C_p	Wind Turbine	η_e (%)	η_m (%)	Probable Energy Production (kWh)
Cam14	6.51	0.4188	Acciona AW70/1500 III	39	96	1,089,556.34
Cam4	6.65	0.4188	Acciona AW70/1500 III	39	96	870,467.93
Cam13	7.15	0.4457	Acciona AW70/1500	39	96	880,222.72
Cam3	7.39	0.4457	Acciona AW70/1500	39	96	1,266,250.78
Cam12	7.47	0.4457	Acciona AW70/1500	39	96	1,330,923.23
Tab3	6.70	0.4188	Acciona AW70/1500 III	39	96	901,712.57
Tab2	6.90	0.4188	Acciona AW70/1500 III	39	96	975,917.27
Tab4	6.96	0.4188	Acciona AW70/1500 III	39	96	995,083.68
Tab1	7.48	0.4457	Acciona AW70/1500	39	96	1,324,954.98
Tam44	6.67	0.4188	Acciona AW70/1500 III	39	96	874,886.55
Tam5	6.72	0.4188	Acciona AW70/1500 III	39	96	892,710.52
Tam57	6.74	0.4188	Acciona AW70/1500 III	39	96	892,710.52
Tam19	6.75	0.4188	Acciona AW70/1500 III	39	96	901,712.57
Tam38	6.81	0.4188	Acciona AW70/1500 III	39	96	933,696.07
Tam32	6.94	0.4188	Acciona AW70/1500 III	39	96	985,469.41
Tam51	6.99	0.4188	Acciona AW70/1500 III	39	96	995,083.68
Tam1	7.01	0.4457	Acciona AW70/1500	39	96	1,084,869.22
Tam13	7.13	0.4457	Acciona AW70/1500	39	96	1,132,490.97
Tam3	7.13	0.4457	Acciona AW70/1500	39	96	1,137,866.71
Tam2	7.17	0.4457	Acciona AW70/1500	39	96	1,159,539.79
Tam9	7.33	0.4457	Acciona AW70/1500	39	96	1,231,874.84
Tam33	7.42	0.4457	Acciona AW70/1500	39	96	1,289,519.75
Tam6	7.44	0.4457	Acciona AW70/1500	39	96	1,295,381.19
Tam4	7.46	0.4457	Acciona AW70/1500	39	96	1,313,072.04
Tam11	7.51	0.4457	Acciona AW70/1500	39	96	1,342,913.46
Tam10	7.52	0.4457	Acciona AW70/1500	39	96	1,348,935.48
Tam7	7.52	0.4457	Acciona AW70/1500	39	96	1,348,935.48
Tam45	7.61	0.4457	Acciona AW70/1500	39	96	1,379,315.79
Tam15	7.65	0.4457	Acciona AW70/1500	39	96	1,416,370.08
Tam26	7.65	0.4457	Acciona AW70/1500	39	96	1,403,945.85
Tam20	7.67	0.4457	Acciona AW70/1500	39	96	1,422,609.58
Tam14	7.68	0.4457	Acciona AW70/1500	39	96	1,428,867.39

Table 3. *Cont.*

ID Point	c (m/s)	C_p	Wind Turbine	η_e (%)	η_m (%)	Probable Energy Production (kWh)
Tam21	7.77	0.4382	Vestas v90 1.8 MW	39.7	96.1	2,404,698.21
Tam27	7.79	0.4382	Vestas v90 1.8 MW	39.7	96.1	2,425,608.51
Tam17	7.84	0.4382	Vestas v90 1.8 MW	39.7	96.1	2,478,413.83
Tam16	7.85	0.4382	Vestas v90 1.8 MW	39.7	96.1	2,489,066.02
Tam22	8.07	0.4382	Vestas v90 1.8 MW	39.7	96.1	2,697,298.23
Tam29	8.08	0.4382	Vestas v90 1.8 MW	39.7	96.1	2,686,059.50
Tam39	8.08	0.4382	Vestas v90 1.8 MW	39.7	96.1	2,686,059.50
Tam23	8.14	0.4382	Vestas v90 1.8 MW	39.7	96.1	2,765,389.54
Tam28	8.20	0.4382	Vestas v90 1.8 MW	39.7	96.1	2,822,999.99
Tam18	8.21	0.4382	Vestas v90 1.8 MW	39.7	96.1	2,834,617.26
Tam34	8.26	0.4382	Vestas v90 1.8 MW	39.7	96.1	2,869,660.21
Ver2	6.46	0.4188	Acciona AW70/1500 III	39	96	345,346.79
Ver29	6.46	0.4188	Acciona AW70/1500 III	39	96	785,191.78
Ver22	6.47	0.4188	Acciona AW70/1500 III	39	96	789,317.12
Ver13	6.60	0.4188	Acciona AW70/1500 III	39	96	844,268.34
Ver25	6.64	0.4188	Acciona AW70/1500 III	39	96	857,301.42
Ver9	6.74	0.4188	Acciona AW70/1500 III	39	96	892,710.52
Ver5	6.82	0.4188	Acciona AW70/1500 III	39	96	924,482.03
Ver6	6.83	0.4188	Acciona AW70/1500 III	39	96	929,081.43
Ver24	6.88	0.4188	Acciona AW70/1500 III	39	96	952,307.41
Yuc13	6.64	0.4188	Acciona AW70/1500 III	39	96	870,467.93
Yuc7	7.27	0.4457	Acciona AW70/1500	39	96	1,231,874.84
Yuc12	7.28	0.4457	Acciona AW70/1500	39	96	1,231,874.84
Yuc3	7.44	0.4457	Acciona AW70/1500	39	96	1,319,004.59
Yuc1	7.54	0.4457	Acciona AW70/1500	39	96	1,373,203.61
Yuc11	7.88	0.4382	Vestas V90 1.8 MW	39	96	1,531,691.62

A total of 58 points were fitted with an appropriate wind turbine C_p, where the wind speed of the site was related to the wind speed at the maximum wind turbine efficiency; with this information, the best option for the wind turbine to be used could be determined based on its C_p. The probable energy production was calculated for each wind turbine selected using Equation (16) based on the electrical and mechanical efficiency given by the manufacturer.

4. Conclusions

The seasonal characterization of wind speeds along the Mexican states of the Gulf of Mexico was done for 141 locations with MERRA-2 data, with records between 1980–2017. An average year of wind data for each location was obtained and these were divided according to the seasons of the year (spring, summer, autumn, and winter). This distinction allowed us to describe the wind characteristics more accurately, especially the WPD, which allowed for establishing wind seasons properly, as well as the description of the prevailing wind direction. It is was found that there were different wind seasons along the states of the Gulf of Mexico and they were established for each state. The wind season in Tamaulipas occurred in winter and spring; Veracruz and Tabasco had a wind season in autumn, winter, and spring; and in Campeche and Yucatan, the wind season was in winter and spring, where these states shared their coast with both the Gulf of Mexico and the Caribbean Sea. This variation allowed us to determine that the wind season could depend upon other factors regarding the climatology of each state.

The state with the greatest wind resource was Tamaulipas, as seen in Figure 9; it had plenty of locations with good values for wind, scale, and shape parameters, where its highest v_m, c, and k was 7.34 m/s, 8.26 m/s, and 2.91 respectively. North and south Veracruz had the highest values of wind speed, where its southern zone corresponded to the Isthmus of Tehuantepec, one of the zones with the greatest wind energy resource in the world; its highest parameters were v_m = 6.10 m/s, c = 6.88 m/s, and $k = 2.41$. Tabasco presented the following parameters: $v_m = 6.67$ m/s, $c = 7.48$ m/s, and $k = 3.07$. These three states (Tamaulipas, Veracruz, and Tabasco) presented the same wind season.

The highest parameters for Campeche were $v_m = 6.57$ m/s, $c = 7.39$ m/s, and $k = 2.71$, and the Yucatan parameters were $v_m = 7.88$ m/s, $c = 7.04$ m/s, and $k = 3.07$; these two states presented the same wind season.

Three clusters were found, which were useful for determining the wind speed in monthly groups and to visualize the wind stations in the study area of the Gulf of Mexico.

Using C_p, optimal wind turbines were determined by comparing it with the parameter c of Weibull distribution. In this study, 58 sites were fitted with a wind turbine C_p, and by analyzing these data, three wind turbines were selected: Acciona AW70/1500 Class III with a $C_p = 0.4188$ at 6.5 m/s, Acciona AW70/1500 Class II with a $C_p = 0.4457$ at 7.5 m/s, and Vestas V90 1.8 MW with a $C_p = 0.4381$ at 8 m/s.

At the end of this study, the probable energy production was calculated for each wind turbine applied at each site with its own respective conditions.

Regarding future research lines, we propose following the proposal to select the most efficient wind turbine by adding more constraints, such as wind direction, roughness, and orography.

Author Contributions: Conceptualization, A.-J.P.-M., G.A. and Q.H.-E.; methodology, A.-J.P.-M., G.A. and Q.H.-E.; formal analysis, A.-J.P.-M., G.A. and Q.H.-E.; investigation, A.-J.P.-M., G.A. and Q.H.-E.; resources, A.-J.P.-M., G.A. and Q.H.-E.; writing—original draft preparation, A.-J.P.-M., G.A. and Q.H.-E. All authors have read and agreed to the published version of the manuscript.

Funding: This research received no external funding.

Acknowledgments: The authors acknowledge the Mexican "Secretaria de Educacion Publica" for the support to the research group UV-CA-466 "Mecanica Electrica" in the project "Evaluación del potencial eólico para disminuir el pico de demanda eléctrica en zonas tropicales. Caso Coatzacoalcos Veracruz México.".

Conflicts of Interest: The authors declare no conflict of interest.

References

1. IEA. International Energy Agency. 2019. Available online: https://www.iea.org/ (accessed on 14 October 2019).
2. IRENA. International Renewable Energy Agency. 2019. Available online: https://www.irena.org/ (accessed on 15 October 2019).
3. IEA. Renewables, International Energy Agency Renewables. 2018. Available online: https://www.iea.org/renewables (accessed on 24 October 2018).
4. Davis, R.E.; Kalkstein, L. Development of an automated spatial synoptic climatological classification. *Int. J. Clim.* **1990**, *10*, 769–794. [CrossRef]
5. Montoya, F.G.; Manzano-Agugliaro, F.; López-Márquez, S.; Hernández-Escobedo, Q.; Gil, C. Wind turbine selection for wind farm layout using multi-objective evolutionary algorithms. *Expert Syst. Appl.* **2014**, *41*, 6585–6595. [CrossRef]
6. Ley de Transición Energética, LTE. 2015. Available online: http://www.diputados.gob.mx/LeyesBiblio/pdf/LTE.pdf. (accessed on 22 October 2018).
7. SIE. Sistema de Información Energética. 2019. Available online: http://sie.energia.gob.mx/bdiController.do?action=cuadro&cvecua=IE0C01 (accessed on 14 October 2019).
8. Bandi, M.M.; Apt, J. Variability of the Wind Turbine Power Curve. *Appl. Sci.* **2016**, *6*, 262. [CrossRef]
9. Hernandez-Escobedo, Q. Wind Energy Assessment for Small Urban Communities in the Baja California Peninsula, Mexico. *Energies* **2016**, *9*, 805. [CrossRef]
10. Avila-Calero, S. Contesting energy transitions: Wind power and conflicts in the Isthmus of Tehuantepec. *J. Polit. Ecol.* **2017**, *24*, 992–1012. [CrossRef]
11. Ospino-Castro, A.; Pena-Gallardo, R.; Hernandez-Rodriguez, A.; Segundo-Ramirez, J.; Munoz-Maldonado, Y.A. Techno-economic evaluation of a grid-connected hybrid PV-Wind power generation system in San Luis Potosi, Mexico. In Proceedings of the 2017 IEEE International Autumn Meeting on Power, Electronics and Computing (ROPEC), Ixtapa, Mexico, 8–10 November 2017.
12. Santamaria-Bonfil, G.; Reyes-Ballesteros, A.; Gershenson, C. Wind speed forecasting for wind farms: A method based on support vector regression. *Renew. Energy* **2016**, *85*, 790–809. [CrossRef]

13. Hernandez-Escobedo, Q.; Manzano-Agugliaro, F.; Zapata-Sierra, A. The wind power of Mexico. *Renew. Sustain. Energy Rev.* **2010**, *14*, 2830–2840. [CrossRef]
14. Mazzeo, D.; Oliveti, G.; Labonia, E. Estimation of wind speed probability density function using a mixture of two truncated normal distributions. *Renew. Energy* **2018**, *115*, 1260–1280. [CrossRef]
15. Gugliani, G.K.; Sarkar, A.; Ley, C.; Mandal, S. New methods to assess wind resources in terms of wind speed, load, power and direction. *Renew. Energy* **2018**, *129*, 168–182. [CrossRef]
16. Green, M.C.; Myrup, L.O.; Flocchini, R.G. A method for classification of wind field patterns and its application to southern California. *Int. J. Climatol.* **1992**, *12*, 111–135. [CrossRef]
17. Jimenez, P.A.; Gonzalez-Rouco, J.F.; Montavez, J.P.; Navarro, J.; Garcia-Bustamante, E.; Valero, F. Surface wind regionalization in complex terrain. *J. Appl. Meteorol. Climatol.* **2008**, *47*, 308–325. [CrossRef]
18. Hernandez-Escobedo, Q.; Manzano-Agugliaro, F.; Gazquez-Parra, J.A.; Zapata-Sierra, A. Is the wind a periodical phenomenon? The case of Mexico. *Renew. Sustain. Energy Rev.* **2011**, *15*, 721–728. [CrossRef]
19. Jimenez, P.A.; Gonzalez-Rouco, J.F.; Garcia-Bustamante, E.; Navarro, J.; Montavez, J.P.; Valero, F.; Arellano, J.V.; Dudhia, J.; Munoz-Roldan, A. Surface wind regionalization over complex terrain: Evaluation and analysis of a high-resolution WRF simulation. *J. Appl. Meteorol. Climatol.* **2010**, *49*, 268–287. [CrossRef]
20. Zapata-Sierra, A.J.; Cama-Pinto, A.; Montoya, F.G.; Alcayde, A.; Manzano-Agugliaro, F. Wind missing data arrangement using wavelet based techniques for getting maximum likelihood. *Energy Convers. Manag.* **2019**, *185*, 552–561. [CrossRef]
21. Kaufmann, P.; Whiteman, C.D. Cluster-Analysis Classification of wintertime wind patterns in the Grand Canyon region. *J. Appl. Meteorol.* **1999**, *38*, 1131–1147. [CrossRef]
22. Jinsol, K.; Hyun-Goo, K.; Hyeong-Dong, P. Surface Wind Regionalization Based on Similarity of Time-series Wind Vectors. *Asian J. Atmos. Environ.* **2016**, *10*, 80–89.
23. Weber, R.O.; Kaufmann, P. Automated classification scheme for wind fields. *J. Appl. Meteorol.* **1995**, *34*, 1133–1141. [CrossRef]
24. Saleh, H.; Abou El-Azm Aly, A.; Abdel-Hady, S. Assessment of different methods used to estimate Weibull distribution parameters for wind speed in Zafarana wind farm, Suez Gulf, Egypt. *Energy* **2012**, *44*, 710–719. [CrossRef]
25. Aukitino, T.; Khan, M.G.M.; Ahmed, M.R. Wind energy resource assessment for Kiribati with a comparison of different methods of determining Weibull parameters. *Energy Convers. Manag.* **2017**, *151*, 641–660. [CrossRef]
26. Akdağ, S.A.; Dinler, A. A new method to estimate Weibull parameters for wind energy applications. *Energy Convers. Manag.* **2009**, *50*, 1761–1766. [CrossRef]
27. Baseer, M.A.; Meyer, J.P.; Rehman, S.; Alam, M.M. Wind power characteristics of seven data collection sites in Jubail, Saudi Arabia using Weibull parameters. *Renew. Energy* **2017**, *102*, 35–49. [CrossRef]
28. Ozay, C.; Celiktas, M.S. Statistical analysis of wind speed using two-parameter Weibull distribution in Alaçatı region. *Energy Convers. Manag.* **2016**, *121*, 49–54. [CrossRef]
29. Katinas, V.; Gecevicius, G.; Marciukaitis, M. An investigation of wind power density distribution at location with low and high wind speeds using statistical model. *Appl. Energy* **2018**, *218*, 442–451. [CrossRef]
30. Herrero-Novoa, C.; Pérez, I.A.; Sánchez, M.L.; García, M.Á.; Pardo, N.; Fernández-Duque, B. Wind speed description and power density in northern Spain. *Energy* **2017**, *138*, 967–976. [CrossRef]
31. Faghani, G.H.R.; Ashrafi, Z.N.; Sedaghat, A. Extrapolating wind data at high altitudes with high precision methods for accurate evaluation of wind power density, case study: Center of Iran. *Energy Convers. Manag.* **2018**, *157*, 317–338. [CrossRef]
32. MERRA. Modern-Era Retrospective Analysis for Research and Applications (MERRA). 2018. Available online: https://gmao.gsfc.nasa.gov/reanalysis/MERRA/ (accessed on 10 August 2018).
33. Shang, Y. Subgraph Robustness of Complex Networks under Attacks. *IEEE Trans. Syst. Man Cybern. Syst.* **2019**, *49*, 821–832. [CrossRef]
34. Fagbenle, R.O.; Katende, J.; Ajayi, O.O.; Okeniyi, J.O. Assessment of wind energy potential of two sites in North-East, Nigeria. *Renew. Energy* **2011**, *36*, 1277–1283. [CrossRef]
35. Ozerdem, B.; Turkeli, M. An investigation of wind characteristics on the campus of Izmir Institute of Technology, Turkey. *Renew. Energy* **2003**, *28*, 1013–1027. [CrossRef]
36. Hocaoğlu, F.O.; Fidan, M.; Gerek, Ö.N. Mycielski approach for wind speed prediction. *Energy Convers. Manag.* **2009**, *50*, 1436–1443. [CrossRef]

37. Carta, J.A.; Ramírez, P.; Velázquez, S. A review of wind speed probability distributions used in wind energy analysis: Case studies in the Canary Islands. *Renew. Sustain. Energy Rev.* **2009**, *13*, 933–955. [CrossRef]
38. Celik, A. Assessing the suitability of wind speed probability distribution functions on wind power density. *Renew. Energy* **2003**, *28*, 1563–1574. [CrossRef]
39. Seguro, J.V.; Lambert, T.W. Modern estimation of the parameters of the Weibull wind speed distribution for wind energy analysis. *J. Wind Eng. Ind. Aerodyn.* **2000**, *85*, 75–84. [CrossRef]
40. Montgomery, D. *Introduction to Statistical Quality Control*, 6th ed.; John Wiley and Sons: Hoboken, NJ, USA, 2009.
41. Montgomery, D.; Runger, G. *Applied Statistics and Probability for Engineers*, 3rd ed.; John Wiley and Sons: Hoboken, NJ, USA, 2003.
42. Vasquez, A.; Sousa, T. Stability Analysis of Distribution Networks using The Cespedes Load Flow Considering Wind Energy Conversion Systems. In Proceedings of the 2018 IEEE PES Transmission & Distribution Conference and Exhibition Latin America, Lima, Perú, 18–21 September 2018.
43. Bilir, L.; İmir, M.; Devrim, Y.; Albostan, A. Seasonal and yearly wind speed distribution and wind power density analysis based on Weibull distribution function. *Int. J. Hydrog. Energy* **2015**, *40*, 15301–15310. [CrossRef]
44. R Core Team. R: A Language and Environment for Statistical Computing. R Foundation for Statistical Computing: Vienna, Austria, 2018. Available online: https://www.R-project.org/ (accessed on 2 January 2019).
45. Grolemund, G.; Wickham, H. Dates and Times Made Easy with lubridate. *J. Stat. Softw.* **2011**, *40*, 1–25. [CrossRef]
46. Neuwirth, E. RColorBrewer: ColorBrewer Palettes. R Package Version 1.1-2. 2014. Available online: https://rdrr.io/cran/RColorBrewer/ (accessed on 3 January 2019).
47. Wickham, H. *ggplot2: Elegant Graphics for Data Analysis, First*; Springer: New York, NY, USA, 2016.
48. R Project. 2019. Available online: https://cran.r-project.org/web/packages/gridExtra/index.html (accessed on 3 January 2019).
49. Kassambara, A. *Practical Guide to Cluster Analysis in R*, 1st ed.; STHDA: Scotts Valley, CA, USA, 2017; Available online: https://www.datanovia.com/en/wp-content/uploads/dn-tutorials/book-preview/clustering_en_preview.pdf (accessed on 25 March 2019).
50. Sugar, C.A.; James, G.M. Finding the number of clusters in data set: An information theoretic approach. *J. Am. Stat. Assoc.* **2003**, *98*, 750–763. [CrossRef]
51. Hao, Y.; Dong, L.; Liao, X.; Liang, J.; Wang, L.; Wang, B. A novel clustering algorithm based on mathematical morphology for wind power generation prediction. *Renew. Energy* **2019**, *136*, 572–585. [CrossRef]
52. Zhu, X.; Liu, W.; Zhang, J. Probabilistic load flow method considering large-scale wind power integration. *J. Mod. Power Syst. Clean Energy* **2019**, *7*, 813–825.
53. van Vuuren, C.Y.J.; Vermeulen, H.J. Wind resource clustering based on statistical Weibull characteristics. *Wind Eng.* **2019**, *43*, 359–376. [CrossRef]
54. Demšar, J.; Curk, T.; Erjavec, A.; Gorup, Č.; Hočevar, T.; Milutinovič, M.; Štajdohar, M. Orange: Data Mining Toolbox in Python. *J. Mach. Learn. Res.* **2013**, *14*, 2349–2353.
55. Wang, K.; Qi, X.; Liu, H.; Song, J. Deep belief network based k-means cluster approach for short-term wind power forecasting. *Energy* **2018**, *165*, 840–852. [CrossRef]
56. Wind Turbines Integration with Storage Devices: Modelling and Control Strategies. Available online: https://www.researchgate.net/publication/221911678_Wind_Turbines_Integration_with_Storage_Devices_Modelling_and_Control_Strategies/link/0deec521b17b05c5ba000000/download (accessed on 1 December 2019).
57. Song, D.; Liu, J.; Yang, J.; Su, M.; Yang, S.; Yang, X.; Joo, Y.H. Multi-objective energy-cost design optimization for variable-speed wind turbine at high-altitude sites. *Energy Convers. Manag.* **2019**, *196*, 513–524. [CrossRef]
58. Elsevier. *Wind Energy a Handbook for Onshore and Offshore Wind Turbines*; Academic Press: Cambridge, MA, USA, 2017; ISBN 978-0-12-809451-8.
59. Ouammi, A.; Dagdougui, H.; Sacile, R. Optimal Planning with Technology Selection for Wind Power Plants in Power Distribution Networks. *IEEE Syst. J.* **2019**, *13*, 3059–3069. [CrossRef]

© 2019 by the authors. Licensee MDPI, Basel, Switzerland. This article is an open access article distributed under the terms and conditions of the Creative Commons Attribution (CC BY) license (http://creativecommons.org/licenses/by/4.0/).

Article

Energy Recovery from Waste Tires Using Pyrolysis: Palestine as Case of Study

Ramez Abdallah [1], Adel Juaidi [1], Mahmoud Assad [1], Tareq Salameh [2] and Francisco Manzano-Agugliaro [3,*]

1 Mechanical Engineering Department, Faculty of Engineering & Information Technology, An-Najah National University, P.O. Box 7, Nablus 00970, Palestine; ramezkhaldi@najah.edu (R.A.); adel@najah.edu (A.J.); m_assad@najah.edu (M.A.)

2 Department of Sustainable and Renewable Energy Engineering, College of Engineering, University of Sharjah, Sharjah 27272, UAE; tsalameh@sharjah.ac.ae

3 Department of Engineering, ceiA3, University of Almeria, 04120 Almeria, Spain

* Correspondence: fmanzano@ual.es

Received: 22 February 2020; Accepted: 3 April 2020; Published: 9 April 2020

Abstract: The first industrial-scale pyrolysis plant for solid tire wastes has been installed in Jenin, northern of the West Bank in Palestine, to dispose of the enormous solid tire wastes in the north of West Bank. The disposable process is an environmentally friendly process and it converts tires into useful products, which could reduce the fuel crisis in Palestine. The gravimetric analysis of tire waste pyrolysis products from the pyrolysis plant working at the optimum conditions is: tire pyrolysis oil (TPO): 45%, pyrolysis carbon black (PCB): 35%, pyrolysis gas (Pyro-Gas): 10% and steel wire: 10%. These results are depending on the tire type and size. It has been found that the produced pyrolysis oil has a High Heating Value (HHV), with a range of 42 – 43 (MJ/kg), which could make it useful as a replacement for conventional liquid fuels. The main disadvantage of using the TPO as fuel is its strong acrid smell and its low flash point, as compared with the other conventional liquid fuels. The produced pyrolysis carbon black also has a High Heating Value (HHV) of about 29 (MJ/kg), which could also encourage its usage as a solid fuel. Carbon black could also be used as activated carbon, printers' ink, etc. The pyrolysis gas (Pyro-Gas) obtained from waste tires mainly consist of light hydrocarbons. The concentration of H_2 has a range of 30% to 40% in volume and it has a high calorific value (approximately 31 MJ/m^3), which can meet the process requirement of energy. On the other hand, it is necessary to clean gas before the burning process to remove H_2S from Pyro-Gas, and hence, reduce the acid rain problem. However, for the current plant, some recommendations should be followed for more comfortable operation and safer environment work conditions.

Keywords: pyrolysis; solid tire wastes; PCB; TPO; Pyro-Gas; industrial scale; Palestine

1. Introduction

The world is facing a huge energy gap due to the increase of the human population and the unexpected growth in industry around the world. In Palestine, the consumption of energy has hugely increased in the last years. According to the Palestinian Central Bureau of Statistics, Palestinians consumed more than 30 to 40 thousand barrels per day of petroleum in 2017 [1]. However, Palestine suffers from a scarcity of natural resources and mineral wealth. The suffering of the Palestinians mostly lies in the scarcity of conventional energy sources, such as oil and gas, and additionally, the expensive price of energy, which is equivalent to the most expensive cities in the world. Inspired by this suffering, Palestinians seek to overcome this dilemma and to find alternative sources for conventional fuels.

Nowadays, the trend in the technology of energy is heading towards alternative sources and mainly, renewable energy sources, in order to overcome the rising demands of energy, which are essential in diverse fields of life. There are many sources for alternative fuels, which are based on recycling industrial and household wastes. These wastes exist in different forms: liquid wastes (e.g., cooking oil) and solid wastes. The solid wastes are any kind of rubbish, tires, paper, cardboard, plastic, wood, glass, metallic wastes and ashes.

The world in general, and Palestine in particular, are facing the problem of solid waste due to the highly increasing numbers of the population, and the lack of material resources needed to solve this problem and reduce its risks. Therefore, the solid waste is randomly collected and assembled in the dumps, contrary to the conditions of environmental health, but this is still not a suitable solution. The waste to energy process is a very simple concept involving recovering the maximum energy from waste with minimum damage to the environment by different thermal processes, such as the pyrolysis process [2]. With the rapid development of the auto industry, "black pollution" is being created by more and more waste tires, which now threaten the living environment of humans.

Millions of tons of waste tires are produced worldwide: about 2.5 million tons is generated in North America, Japan generates one million tons and the European Community generates more than 2.5 million tons [3]. In Palestine, in 2012, it was estimated that more than 380 thousand scrap tires were generated from different sources [4]. The tires are composed of vulcanized rubbers, which include natural rubber (NR), styrene butadiene (SBR), and polybutadiene (BR), carbon black, steel wire, cords and additives.

Getting rid of non-biodegradable solid waste tires is one of the most important environmental problems in the world. The durability and immunity of these wastes towards biological degradation makes disposal and re-treatment difficult [2,5]. In Palestine, the solid waste tires are randomly dumped in landfills. These landfills usually face accidental fires that produce extremely thick black smoke and toxic emissions. Such emissions are deleterious to health and may cause the acid rain [5]. Therefore, the biomass pyrolysis is a suitable solution to recycle these waste tires with more environmentally friendly methods. Pyrolysis is considered a cheap way to obtain alternative liquid fuel. By 2050, the International Energy Agency expects that 27% of the fuel used in the transportation industry may be sustainably produced from biomass and waste resources [6].

Tire recycling is the conversion of scrap rubber tires into hydrocarbons for reuse as oils and fuels, and the recovery of steel and carbon is one of the more desirable methods for economically reclaiming scrap tires. Tire recycling is achieved by the pyrolysis process, see Figure 1.

Pyrolysis is the thermochemical decomposition of tires at high temperatures without oxygen. Pyrolysis takes place under low pressure and high temperatures above 420 °C. Practically, pyrolysis cannot be performed oxygen free, since with a limited quantity of oxygen present in the reactor, oxidation will occur at a limited quantity. Pyrolysis of waste tires will produce pyrolysis black carbon (PCB) (30–35 wt%), crude tire pyrolysis oil (TPO) (40–45 wt%), scrap steel (10–15 wt%) and pyrolysis gas (Pyro-Gas) (10–15 wt%) [7].

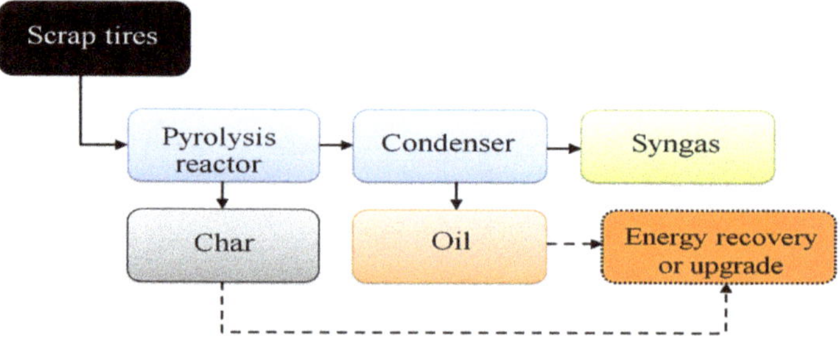

Figure 1. Products of the tire pyrolysis process.

Pyrolysis Technology and Energy Security in Palestine

Lately, increasing energy consumption, high-price volatility in fossil fuels, interruptions in energy imports and exports, climate change and pollution of the environment are just a few of the many energy security issues in countries throughout the world [8]. Renewable energy and bioenergy play a substantial role in economy, social and environmental life [9,10], bringing significant benefits for developing countries, specifically Palestine. Besides, use of these energy sources avoids or eliminates the negative environmental impacts. Regarding the political situation in Palestine and to avoid the highly frequent blackouts, use of renewable energy and bioenergy (such as pyrolysis) will be a sustainable solution to guarantee the energy supply in the long term. A significant driver of pyrolysis is also the growing interest in environmental sustainability [11]. Energy security is a significant factor that enables the system to function optimally and sustainably, free of threats and risks [12]. Energy security guarantees energy supply reliability at an affordable price, and is therefore affected by economic, political and technical conditions [13].

Energy supply security and environmental issues are considered to be among the main concerns for most countries [8]. While Palestine is highly dependent on imported energy, the results show that solar energy and bioenergy (Pyrolysis) are the most promising renewable and sustainable energy sources. The advantages of pyrolysis include a highly flexible technology that can handle a wide variety of feedstocks consisting of forest residues, food waste, tire waste and solid municipal waste [11]. Palestine is among the most promising locations for utilization of bioenergy and renewable energy sources because of the geographic location of this area, which supplies it with plenty of high solar radiation, and the availability of bioenergy sources. Also, it is highly dependent on energy imported from Israel [9].

To achieve sustainability, this project involves producing a renewable fuel from the pyrolysis process of waste tires. The fuel production from this process involves less costs and less greenhouse gas emissions and makes the environment cleaner. The sustainability aims of this project are directed at solving three main issues: economic, environment and social, in order to achieve energy security and a clean environment in Palestine.

The present study aims at solving the firm technical problems of tire wastes in Palestine. The novelty of this research is that it adds sustainability as an aspect of energy security in Palestine's energy systems. This is done by introducing the first commercial plant of tire pyrolysis in Palestine.

2. Literature Review

The disposal of waste tires is considered a main economic and environmental problem in the waste management field. For the time being, attention is growing in pyrolysis due to its ability to make products of economic value in different fractions by changing the operating conditions, such as the operating temperatures and heating rate [14]. Pyrolysis also transforms waste tires into tire pyrolysis oil (TPO), pyrolysis carbon black (PCB) and pyrolysis gas (Pyro-Gas), which are considered as high value fuels and chemicals [15,16]. Table 1 below summarizes the pyrolysis research during the period from 1980 to 2018.

Table 1. Pyrolysis research for different countries worldwide: A representative list of waste tire pyrolysis processes.

Year	Country	Description/Module	Remarks	References
1980	Japan	In 1979, the Japanese company Sumitomo Cement Co., Ltd., established a pyrolysis plant with a rotary kiln reactor with an annual capacity of 7000 tons in AKO City, Hyogo Prefecture.	The pilot scale plant was turned into a commercial plant with low costs to recover fuel oil and carbon black.	[17,18]
1996	France	A pilot scale batch reactor was designed and used to investigate the effects of different operating conditions like temperature, fuel-to-air ratios and reaction times.	It was noticed that there were low percentages of emissions of SO_2 and NOx, although the materials relatively contained high-sulfur materials.	[19]
1997	Italy	The pyrolysis process is experimentally investigated using a pilot scale reactor.	A number of gases were detected in the produced gas, such as H_2, O_2, N_2, CO and CH_4.	[20]
2000	China	This review summarizes the research about pyrolysis of waste tires in china focusing mainly on the mechanism of the pyrolysis process and the pyrolysis reactor design.	The pyrolysis industry in China has the tendency towards a small scale and low initial cost pyrolysis plant.	[21]
2004	China	The pyrolysis process is studied using a pilot scale rotary kiln reactor. The effects of operating temperatures are investigated at a range between 450 and 650 °C.	A rotary kiln is considered an excellent alternative for pyrolysis of waste tires.	[22]
2005	Spain	A comparison is made between a fixed-bed laboratory scale and a rotatory pilot scale reactor at the same operating conditions to obtain black carbon, liquid fuel, and Pyro-Gas in the two scales.	No significant differences were found between the pyrolysis black carbon obtained from tires in the two scales.	[23]
2005	China	Waste tires were subjected to a pyrolysis process using a rotary kiln pilot scale reactor.	Raw pyrolysis black carbon and activated black carbon have acceptable ability for adsorption of methylene-blue compared to commercial black carbon.	[24]
2007	Bangladesh	Heavy automotive waste tires were subjected to a pyrolysis process in a fixed-bed reactor under deferent process parameters.	Pyrolysis tire oil obtained have fuel properties that make it comparable to liquid petroleum fuels.	[25]
2008	Japan	A fixed-bed reactor that heated using a fire tube was used to pyrolyze the car tire wastes under a nitrogen condition and it was tested under different pyrolysis conditions.	The maximum liquid yield is obtained at a temperature of 475 °C.	[26]
2008	Bangladesh	The tire pyrolysis oils produced were subjected to elemental analysis, and chromatographic and spectroscopic techniques.	The tire pyrolysis oil could be used as fuel with a high heating value of 42 MJ/kg.	[27]
2008	Bangladesh	Bicycle waste tires were subjected to a pyrolysis process in a fixed-bed reactor under several process parameters.	The pyrolysis tire oil obtained could be comparable to fuels derivative from crude oil if the pyrolysis is performed under suitable operating conditions.	[28]
2009	Bangladesh	Waste tires were subjected to thermogravimetric analysis at temperatures between 30 and 800 °C and heating rates of 10 and 60 °C/min.	The pyrolysis of waste tires is significantly affected by the heating rate.	[29]

Table 1. *Cont.*

Year	Country	Description/Module	Remarks	References
2010	Bangladesh	Pyrolysis technology using a new heating system to dispose of the solid waste tire and to recovery liquefied fuels.	The pyrolysis process is considered an environmentally friendly solution for the waste tires issue.	[30]
2010	Italy	The results of several experiments were used to develop a numerical model using the commercial code ChemCAD.	The results of an experimental and numerical study for gasification in a rotary kiln reactor of tires were explained.	[31]
2011	Bangladesh	Feasibility study was performed to obtain liquid oil, black carbon and Pyro-Gas from waste tires	The feasibility study was performed to three different scales. It can be shown that the medium commercial scale is the most economical scale.	[32]
2012	India	A review of pyrolysis of waste tires was made focusing on several operating conditions.	The pyrolysis process has three products with commercial value: black carbon, liquid fuel and Pyro-Gas.	[33]
2013	UK	This review presents that the interest in pyrolysis of waste tires to obtain products with commercial value is growing.	The Pyro-Gas produced from pyrolysis of waste tires consists of H_2, CO, CO_2, H_2S and C_1-C_4 hydrocarbons. This review has focused on upgrading pyrolysis black carbon to activated black carbon and to black carbon with a better quality.	[34]
2013	Bangladesh	A small pyrolysis scale of waste tires could reduce the crisis of liquid fuel in waste in Bangladesh.	The Pyro-Gas produced from the pyrolysis of waste tires has a high heating value of 37 MJ/m^3, which makes it a high enough heat source required to complete the pyrolysis process.	[5]
2013	Bangladesh	A fixed bed pilot scale reactor was used to obtain pyrolysis oil from waste tires	The properties of the pyrolysis oil were comparable to commercial diesel.	[35]
2014	France	The fuel was obtained from the pyrolysis of waste tires at optimum conditions for the temperature, heating rate and inert gas flow rate.	Without using inert gas flow, the optimum temperature for the pyrolysis process is 465 °C.	[36]
2017	Poland	The pyrolysis process for the waste tires was explained using a mathematical model.	This work shows the importance of the heat and mass.	[37]
2017	Poland	Two samples of waste tires and rubber materials were studied and compared at different heating rates (e.g., 10, 20 and 50 K/min).	The thermogravimetry equipment (TG) was used to perform a kinetic study and determine the kinetic mechanism of the process.	[38]
2018	Bangladesh	The pyro oil was produced from pyrolysis of scrap tires using a fixed-bed batch reactor.	The pyro oil produced from this reactor is used as fuel in a boiler and as furnace oil. The char is potentially used as a fertilizer, and it can be used to produce shoes and conveyor belts.	[39]

Pyrolysis Gas: Pyro-Gas. The pyrolysis process allows the communities to have an alternative environmentally friendly source of energy, reduce the need for non-renewable fossil fuels and minimize the landfilling problems.

3. Methodology

As mentioned earlier, this study was conducted in Palestine to solve the environmental problem of direct burning of the tires and reduce pollution. It is believed that the pyrolysis process of such waste is a suitable solution. The classical tires consist of different materials such as synthetic rubber, steel wires and carbon black. These materials are homogenously distributed over the tire volume. To dispose of these materials using pyrolysis technology, the tires are usually shredded to smaller pieces, washed with water and dried. The tires are placed in a rotary reactor to extract the wires and produce gaseous and liquefied oil. Since no oxygen exists inside the reactor, the carbon black and steel properties will not change.

This study investigates the pyrolysis of tires using an industrial-scale system. The industrial scale is basically a commercial pyrolysis plant, on about 6000 m^2 of land in the north of Jenin City, West Bank, Palestine. The pyrolysis plant consists of five sections: the main plant section, the offices section, the laborers rooms section, the waste tire section and the warehouse for the products from pyrolysis. This plant was established in 2013 to get rid of the enormous tire wastes in the north of West Bank. The flow diagram of the pyrolysis process system in this plant is shown in Figure 2.

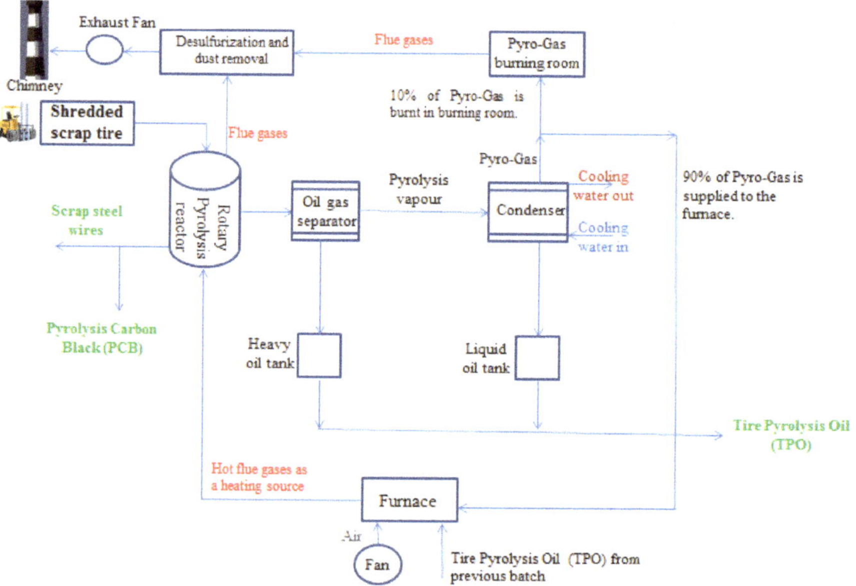

Figure 2. Flow diagram for the tire pyrolysis process.

The plant consists of a horizontal axis cylindrical reactor (6 m long and 2 m diameter) that has a batch mode and rotates with a rotational speed of 0.4 rpm. The reactor chamber can fit up to five tons of partly shredded tires (only small side rings are removed, and steel wires are kept inside). The cylindrical chamber is made from steel and it is initially heated by flames from external source power. Once the system starts producing gaseous fuel from the pyrolysis process, the chamber is now heated by burning the produced gaseous fuel. Therefore, the system is self-sustained in terms of energy. Electricity was used to rotate the chamber for better stirring. The tires were added, and a heavy door was closed tightly. The heat exchangers consist of two groups. The first group contains three heat exchangers that condensate the passing gas coming from the separator. The second group contains two heat exchangers to condensate the remaining gas coming out from the first group. The condensate fuel is filtered to purify the fuel from the contaminations (such as carbon black).

The heating system is composed of an insulated furnace below the reactor. The furnace contains two liquid burners and two gas burners. An insulating cover surrounds the reactor drum. The resulting hot gases pass between the reactor drum and its insulated cover and give enough heat for the pyrolysis process.

Two storage tanks were used in this system: the first tank stores the produced fuel from the first group of heat exchangers, while the second tank stores the remaining produced liquid fuel. The fuel in the first tank is heavier (higher molecular weight fractions) and less clear than the fuel in the second tank. However, the heavy fuel is still demanded in the market for its high heating value and efficient burning. The excess gases that do not liquefy are stored into a third tank, which feeds the heating process. The exhaust gas from the combustion process is filtered for toxic gases before it is spilled out to the atmosphere. The reactor works at a relatively low pressure (the maximum reactor pressure is 0.08 bars). Figure 3 shows the reactor temperatures during heating and cooling processes for four different runs. The total tire load for each run (batch) was 5000 kg. Firstly, the pyrolysis oil produced from previous pyrolysis runs is burned to heat the reactor for the first 3 to 4 h until the temperature reaches 420–500 °C. Secondly, the temperature is held at 420–500 °C by burning the Pyro-Gas produced from the current pyrolysis run. Finally, the reactor is left to cool down naturally for around 10 h. When the reactor is cooled below 60 °C, steel wire and pyrolysis black carbon (PCB) are removed. The PCB fraction is in the form of powder and separated from steel wire while the reactor drum is continuously rotating.

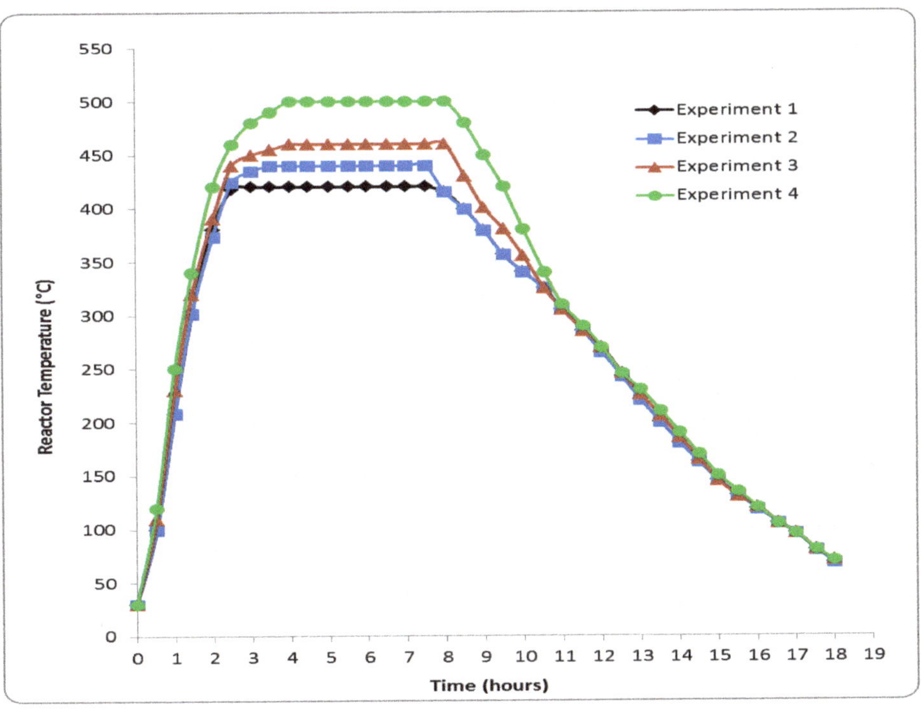

Figure 3. The reactor temperatures during heating and cooling for the four experiments.

The flue exhaust gas system consists of a chimney, a desulfurization and dust removal unit and an exhaust suction fan. The flue gas is passed through the desulfurization and dust removal unit to filter the dust and remove the sulfur from the flue gas. Finally, the filtered flue gas is exhausted to the atmosphere.

Analaytical Techniques

The original sample of tire and the sample of pyrolysis carbon black (PCB) for the four experiments underwent thermogravimetric analysis to take simultaneous results of differential thermogravimetric analysis (DTA) signals. The thermogravimetric analyses are done by burning the samples (approximately 15 mg for each sample) with a constant heating rate and gas flow rate (15 °C/min) and (100 mL/min), respectively. The calorific value, proximate and elemental analyses were performed, using the procedures of American Society for Testing and Materials (ASTM) standards.

The four samples of tire pyrolysis oil (TPO) from different experiments in the industrial pyrolysis plant were homogenized by mixing the oil well before the analysis was carried out. ASTM standard procedures were also used to find the physical properties of tire pyrolysis oil (TPO): flash point, viscosity, pour point, density and HHV. Elemental analysis of TPO was performed by an elementary analyzer.

4. Results and Discussion

4.1. Yields of the Pyrolysis Products

The yields for each of the fractions of the pyrolysis process obtained from the four experiments are shown in Table 2. The total tire load for each experiment (batch) was 5000 kg. The products of the pyrolysis process for the four experiments were observed to be similar with small differences. These differences are attributed to the tires' diversity and their size.

Table 2. Yield of pyrolysis products.

Products	Average (Kg)	Average (%wt)
Crude tire pyrolysis oil (TPO)	2249.8 ± 63.4	45 ± 1.63
Pyrolysis carbon black (PCB)	1708.5 ± 102.1	34 ± 2.45
Scrap steel	547.5 ± 21.3	11 ± 0.82
Pyrolysis gas (Pyro-Gas)	494.3 ± 70.3	10 ± 1.83

4.2. Characterization of Solid Tires and Pyrolysis Carbon Black (PCB)

The original tire and the carbon black fraction for the four experiments were subjected to thermogravimetric analyses. Calorific value and elemental analyses were done according to the ASTM standard procedure.

Proximate and elemental analyses were done on the original tire and the pyrolysis carbon black (PCB) fractions obtained from the four experiments and the results are tabulated in Table 3. It can be concluded that small differences were obtained between the PCB fractions for the four experiments. It can also be noticed that chlorine and sulfur were condensed in the PCB fraction. Approximately, most of the chlorine existing in the waste tires was concentrated in the PCB fraction. Therefore, it is important to take this into consideration once PCB is used as a fuel. So, a system for cleaning the gas would be necessary for the residual gases resulting from combustion of PCB. The PCB residue of the pyrolysis tire process will produce three times the amount of chlorine produced by the original tires [40]. The High Heating Value (HHV) for pyrolysis carbon black (PCB) is about 36 MJ/kg. On the other hand, the HHV for convectional liquid petroleum is 39.5 to 45 MJ/kg. The HHV of PCB is about two times that of wooden biomass (18.5 to 21 MJ/kg). From the analysis results above, it could be concluded that PCB is a possible source of energy.

Figure 4 shows a Derivative Thermogravimetry (DTG) for the original tire and the pyrolysis carbon black (PCB) fractions obtained from the four experiments. It can be concluded from Figure 4 that the highest mass loss rate for the original tire occurs nearly at 390 °C, while for the PCB, it occurs at 570 °C.

Table 3. Proximate, elemental and heating value analysis for the tire and carbon black.

	Original Tire (%wt)	Carbon Black (%wt)
C [1] (%)	84.92	95.42 ± 0.16
H [1] (%)	7.58	0.77 ± 0.20
N [1] (%)	0.51	0.22 ± 0.07
S [1] (%)	2.41	3.29 ± 0.09
Cl [1] (%)	0.05	0.19 ± 0.01
O [1] (%)	4.53	0.12 ± 0.07
Ash	7.51	16.5 5 ± 0.34
Moisture	0.97	1.16 ± 0.14
Volatile matter (%)	62.42	2.50 ± 0.74
HHV (MJ/kg)	35.7	28.70 ± 0.18

[1] Results in dry basis and ash free.

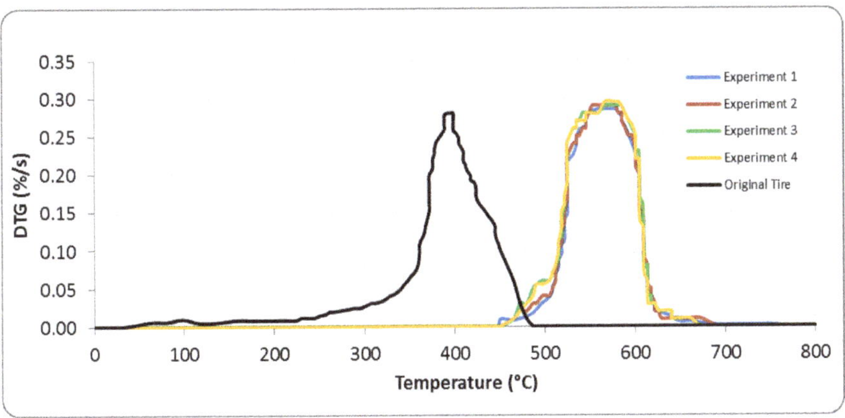

Figure 4. Derivative Thermogravimetry (DTG) plots of the original tire and the pyrolysis carbon black (PCB) fractions obtained from the four experiments.

4.3. Characterization of Tire Pyrolysis Oil (TPO)

The tire pyrolysis oil (TPO) obtained from the four runs of the pyrolysis plant were relatively similar to each other. TPO is a dark brown oil with a strong acrid smell. It interacts with human skin and it usually leaves spots and a bad smell for several days. So, careful treatment is required during use of TPO. The TPO properties for the four experiments are compared to commercial diesel and the results are tabulated in Table 3. It can be concluded that the density of TPO is more than the density of the commercial diesel and less than the density of the heavy fuel oil (985 kg/m^3 at 20 °C). The viscosity of the TPO is more than the viscosity of the commercial diesel and less than the viscosity of heavy fuel oil (200 cSt at 50 °C). Low viscosity of the tire pyrolysis oil is a wanted property in liquid to ease handling and transportation of the fuel. The flash temperature of a fuel gives an indication for the hazards of fire during storage and usage of the fuel. From Table 4, it can be concluded that the flash point of TPO is between 30 and 33 °C. The flash point of TPO is low if it is compared to conventional diesel fuel. The flash point for the conventional diesel fuel is 55 °C, for kerosene the flash point is 23 °C and for light fuel oil it is 79 °C. The flash point of the TPO is low because TPO is a mixture of un-refined liquids that has a wide range of distillation. Table 4 shows that the pour point of the TPO is relatively low when compared with diesel. The pH value of the TPO is between 4.27 and 4.37, which is considered low acidic fuel. This weak acidity is not a problem regarding storage and handling of the TPO, for example, soft drink companies like Cola or Pepsi utilize Polyethylene Terephthalate

(PET) plastic for storage and handling, although the value of pH for their drinks is about 2.5 [41]. Subsequently, TPO should be filtered, centrifuged and desulfurized before it can be utilized as a heavy furnace oil or blended with conventional diesel for furnaces, boilers and power plants.

Table 4. Characteristics of the tire pyrolysis oil (TPO) in comparison to commercial diesel.

Analyses	Commercial Diesel	Tire Pyrolysis Oil (TPO)
Elemental (wt%)		
C	86.50	85.69 ± 0.38
H	13.20	9.20 ± 0.08
C/H	6.52	9.32 ± 0.12
N	65 ppm < 1	0.59 ± 0.06
S	0.11–0.70	1.11 ± 0.06
Ash	0.0	0.13 ± 0.02
O	0.01	3.29 ± 0.36
H/C molar ratio	1.83	1.29 ± 0.02
O/C molar ratio	-	0.029 ± 0.003
Empirical formula	-	$CH_{1.29}O_{0.029}N_{0.006}$
Density (kg/m^3)	820-860	962.5 ± 2.08
Viscosity (cSt)	2.0–4.5 [a]	4.86 [b] ± 0.03
Flash Point (°C)	>55	31.5 ± 1.29
Pour point (°C)	-	−5.4 ± 0.48
Moisture (wt%)	79 ppm	ND
pH value	-	4.32 ± 0.05
HHV (MJ/kg)	45.10	42.46 ± 0.18

ND: Not Determined, [a] at 40 °C, [b] at 30 °C.

4.4. Characterization of the Pyrolysis Gas (Pyro-Gas) Fraction

The compositional analysis showed that the pyrolysis gas (Pyro-Gas) consists mainly of light hydrocarbons and the concentrations of methane and hydrogen were the highest. Hydrogen concentration could reach 40% in volume at high pyrolysis temperatures of 550 °C [40]. Pyro-Gas has a high heating value (38–41 MJ/m^3), which is considered close to the HHV of natural gas (around 39 MJ/m^3), that could meet the need of energy during the pyrolysis process or other industrial applications. However, it would be very important to clean Pyro-Gas before burning to take out the hydrogen sulfide (H_2S) from the Pyro-Gas fraction. The corrosive nature of H_2S would damage the downstream equipment. H_2S amount must be less than 500 ppm in order to be used for commercial alternative engines. In addition, SO_2 emitted after the combustion of the gas fraction should satisfy the environmental constrains.

5. Summary

The gravimetric analysis of tire waste pyrolysis products from the reactor working at the optimum conditions is tire pyrolysis oil (TPO): 45%, pyrolysis carbon black (PCB): 35%, pyrolysis gas (Pyro-Gas): 10% and steel wire: 10%. It can be concluded that chlorine and sulfur were concentrated in the PCB fraction due to the nature of the tires' composition. Approximately all chlorine in the tires was concentrated in the PCB fraction. Therefore, it is very important to take this result into account once PCB is used as a fuel.

The main disadvantage of using the TPO as fuel is its strong acrid smell, which could last for a few days if it interacts with human skin or clothes. The density of TPO is more than that of conventional diesel. However, it is still less than the density of heavy fuel oil. On the other hand, the flash point of TPO is low if it is compared with the other refined fuels, which are considered a disadvantage associated with the hazards of fire during the usage and storage of the fuel.

The pH value of the TPO is considered to be of a weak acidic nature but this weak acidity is not a problem regarding storage and handling of TPO. The compositional analysis showed that the pyrolysis gas (Pyro-Gas) consists mainly of light hydrocarbons and the concentration of methane and hydrogen were the highest. The concentration of H_2 is about 40% in volume and it has a High Calorific

Value (38–41 MJ/m^3), which could meet the process requirement of energy. However, it would be very important to clean Pyro-Gas before burning so as to take out the hydrogen sulfide (H_2S) from Pyro-Gas, and hence reduce the acid rain problem.

The heavy fuel oil price in Palestine is around 0.5 $/liter. However, the production cost of the tire pyrolysis oil (TPO) is around 0.15 $/liter. This cost is less than one third the price of the conventional heavy fuel oil in Palestine. If three pyrolysis plants, such as the one in the present study with a capacity of 330 (tons/month), were founded, they would be able to get rid of the waste tires generated yearly in Palestine (3800 metric tons). If all the waste tires generated in Palestine were pyrolyzed, the annual import heavy fuel oil would decrease by 1700 tons (11,850 barrels) and produce more than 1330 tons of pyrolysis black carbon and 380 tons of steel wire, yearly. In addition, and most importantly, the environmental hazardous problem of the waste tires could be solved in the correct way, reducing the dependence on imported furnace oil and creating new job opportunities.

6. Conclusions and Recommendations

The following recommendations should be applied for more comfortable operation and safer work and environmental conditions:

1. Handling and transportation of the TPO and PCB must be in a closed system.
2. To produce alternative diesel from pyrolysis tire oil (TPO), more filtration, desulfurization and wet scrubber with CaOH column must be added.
3. To make the technology sustainable, better utilization of PCB must be established.
4. The workers in the pyrolysis plant must be skillful and well trained.
5. There is chlorine present and there is a risk of the formation of carcinogenic dioxins. However, the objective of this research is the energy recovery from waste tires using pyrolysis and the analysis of risk of the formation of carcinogenic dioxins can be considered in the next research paper.
6. Finally, it can be summarized that the pyrolysis plant should be operated only under constant monitoring and advisory support by a team specialized in tire pyrolysis, since it is a new technology.

Author Contributions: R.A., M.A. and A.J. conceived and performed the experiments used in the article; A.J., M.A. and R.A. analyzed the data; R.A., A.J., and M.A. wrote the paper. F.M.-A. and T.S. revised and supervised the manuscript. The authors shared the work on the structure and aims of the manuscript, paper drafting, editing and reviewing the manuscript. All authors have read and agreed to the published version of the manuscript.

Funding: This research was funded by An-Najah National University with grant number [ANNU-1819-Sc023].

Acknowledgments: The authors would like to acknowledge the Deanship of Scientific Research at An-Najah National University for their grant which helped in accomplishing the current study.

Conflicts of Interest: The authors declare no conflict of interest.

List of Abbreviations

Abbreviation	Definition
BR	Polybutadiene
NR	Natural Rubber
PCB	Pyrolysis Carbon Black
Pyro-Gas	Pyrolysis Gas
SBR	Styrene Butadiene
TPO	Tire Pyrolysis Oil
PET	Polyethylene Terephthalate
HHV	High Heating Value
ASTM	American Society for Testing and Materials
TGA	Thermogravimetric Analysis
DTG	Derivative Thermogravimetry

References

1. Palestinian Central Bureau of Statistics (PCBS). State of Palestine, Household Energy Survey. Available online: http://www.pcbs.gov.ps/ (accessed on 14 June 2019).
2. Huffman, G.P.; Shah, N. Can waste plastics and tires be recycled economically. *Chemtech* **1998**, *28*, 34–43.
3. Wang, H.; Xu, H.; Xuan, X. Review of waste tire reuse and recycling in China –current situation, problems and countermeasures. *Adv. Nat. Sci.* **2013**, *1*, 81–90.
4. Amer EL-Hamouz. The Development of National Mangle Plan for Hazardous Waste Management for Palestinian National Authority. 2010. Available online: http://environment.pna.ps/ar/files/Part_one_Final_Report_on_The_Development_of_a_National_Master_Plan_for_Hazardous_Waste_Management_for_the_Palestinian_National_Authority_en.pdf (accessed on 29 June 2019).
5. Islam, M.R.; Islam, M.N.; Mustafi, N.N.; Rahim, M.A.; Haniu, H. Thermal recycling of solid tire wastes for alternative liquid fuel: The first commercial step in Bangladesh. *Procedia Eng.* **2013**, *56*, 573–582. [CrossRef]
6. International Energy Agency. *Technology Roadmap: Biofuels for Transport*; International Energy Agency: Paris, France, 2011. [CrossRef]
7. Aydın, H.; İlkılıç, C. Optimization of fuel production from waste vehicle tires by pyrolysis and resembling to diesel fuel by various desulfurization methods. *Fuel* **2012**, *102*, 605–612. [CrossRef]
8. Bekhrad, K.; Aslani, A.; Mazzuca-Sobczuk, T. Energy Security in Andalusia: The Role of Renewable Energy Sources. *Case Stud. Chem. Environ. Eng.* **2019**, 100001. [CrossRef]
9. Juaidi, A.; Montoya, F.G.; Ibrik, I.H.; Manzano-Agugliaro, F. An overview of renewable energy potential in Palestine. *Renew. Sustain. Energy Rev.* **2016**, *65*, 943–960. [CrossRef]
10. Juaidi, A.; Montoya, F.G.; Gázquez, J.A.; Manzano-Agugliaro, F. An overview of energy balance compared to sustainable energy in United Arab Emirates. *Renew. Sustain. Energy Rev.* **2016**, *55*, 1195–1209. [CrossRef]
11. Sharifzadeh, M.; Sadeqzadeh, M.; Guo, M.; Borhani, T.N.; Konda, N.M.; Garcia, M.C.; Wang, L.; Hallett, J.; Shah, N. The multi-scale challenges of biomass fast pyrolysis and bio-oil upgrading: Review of the state of art and future research directions. *Prog. Energy Combust. Sci.* **2019**, *71*, 1–80. [CrossRef]
12. Azzuni, A.; Breyer, C. Energy security and energy storage technologies. *Energy Procedia* **2018**, *155*, 237–258. [CrossRef]
13. Gillessen, B.; Heinrichs, H.; Hake, J.F.; Allelein, H.J. Energy security in context of transforming energy systems: A case study for natural gas transport in Germany. *Energy Procedia* **2019**, *158*, 3339–3345. [CrossRef]
14. Kramer, C.A.; Loloee, R.; Wichman, I.S.; Ghosh, R.N. Time resolved measurements of pyrolysis products from thermoplastic poly-methyl-methacrylate (PMMA). In *ASME 2009 International Mechanical Engineering Congress and Exposition*; American Society of Mechanical Engineers Digital Collection: New York, NY, USA, 2009; pp. 99–105.
15. Chowdhury, R.; Sarkar, A. Reaction kinetics and product distribution of slow pyrolysis of Indian textile wastes. *Int. J. Chem. React. Eng.* **2012**, *10*. [CrossRef]
16. Biswal, B.; Kumar, S.; Singh, R.K. Production of hydrocarbon liquid by thermal pyrolysis of paper cup waste. *J. Waste Manag.* **2013**, *2013*, 1–7. [CrossRef]
17. Kagayama, M.; Igarashi, M.; Fukada, J.; Kunii, D. *Thermal Conversion of Solid Wastes and Biomass*; America Chemical Society: Washington, DC, USA, 1980; Volume 130, pp. 527–531.
18. Kawakami, S.; Inoue, K.; Tanaka, H.; Sakai, T. Pyrolysis Process for Scrap Tires. In *Thermal Conversion of Solid Wastes and Biomass*; Chapter 40; ACS Publications: Washington, DC, USA, 1980; pp. 557–572. Available online: https://pubs.acs.org/doi/pdf/10.1021/bk-1980-0130.ch040 (accessed on 29 August 1980).
19. Avenell, C.S.; Sainz-Diaz, C.I.; Griffiths, A.J. Solid waste pyrolysis in a pilot-scale batch pyrolyser. *Fuel* **1996**, *75*, 1167–1174. [CrossRef]
20. Fortuna, F.; Cornacchia, G.; Mincarini, M.; Sharma, V.K. Pilot-scale experimental pyrolysis plant: Mechanical and operational aspects. *J. Anal. Appl. Pyrolysis* **1997**, *40*, 403–417. [CrossRef]
21. Yongrong, Y.; Jizhong, C.; Guibin, Z. Technical advance on the pyrolysis of used tires in China. *China-Japan International Academic SymposiumEnvironmental Problem in Chinese Iron-Steelmaking Industries and Effective Technology Transfer*; Sendai, Japan, 2000; pp. 84–93. Available online: http://www2.econ.tohoku.ac.jp/~{}kawabata/Omurapro/omuraCDM/symp_jc/10yang.pdf (accessed on 8 April 2020).
22. Li, S.Q.; Yao, Q.; Chi, Y.; Yan, J.H.; Cen, K.F. Pilot-scale pyrolysis of scrap tires in a continuous rotary kiln reactor. *Ind. Eng. Chem. Res.* **2004**, *43*, 5133–5145. [CrossRef]

23. Díez, C.; Sánchez, M.E.; Haxaire, P.; Martínez, O.; Morán, A. Pyrolysis of tyres: A comparison of the results from a fixed-bed laboratory reactor and a pilot plant (rotatory reactor). *J. Anal. Appl. Pyrolysis* **2005**, *74*, 254–258. [CrossRef]
24. Li, S.Q.; Yao, Q.; Wen, S.E.; Chi, Y.; Yan, J.H. Properties of pyrolytic chars and activated carbons derived from pilot-scale pyrolysis of used tires. *J. Air Waste Manag. Assoc.* **2005**, *55*, 1315–1326. [CrossRef]
25. Rofiqul, I.M.; Haniu, H.; Rafiqul, A.B.M. Limonene-rich liquids from pyrolysis of heavy automotive tire wastes. *J. Environ. Eng.* **2007**, *2*, 681–695. [CrossRef]
26. Islam, M.R.; Il, K.S.; Haniu, H.; Beg, M.R.A. Fire-tube heating pyrolysis of car tire wastes: End uses of product liquids as fuels and chemicals. *Int. Energy J.* **2008**, *9*. Available online: http://www.rericjournal.ait.ac.th/index.php/reric/article/view/485 (accessed on 8 April 2020).
27. Islam, M.R.; Haniu, H.; Beg, M.R.A. Liquid fuels and chemicals from pyrolysis of motorcycle tire waste: Product yields, compositions and related properties. *Fuel* **2008**, *87*, 3112–3122.
28. Islam, M.R.; Tushar, M.S.H.K.; Haniu, H. Production of liquid fuels and chemicals from pyrolysis of Bangladeshi bicycle/rickshaw tire wastes. *J. Anal. Appl. Pyrolysis* **2008**, *82*, 96–109. [CrossRef]
29. Islam, M.R.; Haniu, H.; Fardoushi, J. Pyrolysis kinetics behavior of solid tire wastes available in Bangladesh. *Waste Manag.* **2009**, *29*, 668–677. [CrossRef] [PubMed]
30. Islam, M.R.; Parveen, M.; Haniu, H.; Sarker, M.I. Innovation in pyrolysis technology for management of scrap tire: A solution of Energy and Environment. *Int. J. Environ. Sci. Dev.* **2010**, *1*, 89. [CrossRef]
31. Donatelli, A.; Iovane, P.; Molino, A. High energy syngas production by waste tyres steam gasification in a rotary kiln pilot plant. Experimental and numerical investigations. *Fuel* **2010**, *89*, 2721–2728. [CrossRef]
32. Islam, M.R.; Joardder, M.U.H.; Hasan, S.M.; Takai, K.; Haniu, H. Feasibility study for thermal treatment of solid tire wastes in Bangladesh by using pyrolysis technology. *Waste Manag.* **2011**, *31*, 2142–2149. [CrossRef]
33. Karthikeyan, S.; Sathiskumar, C.; Moorthy, R.S. Effect of process parameters on tire pyrolysis: A review. *J. Sci. Ind. Res.* **2012**, *71*, 309–315.
34. Williams, P.T. Pyrolysis of waste tyres: A review. *Waste Manag.* **2013**, *33*, 1714–1728. [CrossRef]
35. Kader, M.; Islam, M.; Hossain, M.; Haniu, H. Development of a Pilot Scale Pyrolysis Plant for Production of Liquid Fuel from Waste Tire. *Mech. Eng. Res. J.* **2013**, *9*, 54–59.
36. Alkhatib, R. Development of an Alternative Fuel from Waste of Used Tires by Pyrolysis. Ph.D. Thesis, L'Université Nantes Angers Le Mans France, Nantes, France, 2014. Available online: https://tel.archives-ouvertes.fr/tel-01186556/ (accessed on 5 August 2019).
37. Rudniak, L.; Machniewski, P.M. Modelling and experimental investigation of waste tyre pyrolysis process in a laboratory reactor. *Chem. Process Eng.* **2017**, *38*, 445–454. [CrossRef]
38. Januszewicz, K.; Klein, M.; Klugmann-Radziemska, E.; Kardas, D. Thermogravimetric analysis/pyrolysis of used tyres and waste rubber. *Physicochem. Probl. Miner. Process.* **2017**, *53*, 802–811.
39. Aziz, M.A.; Rahman, M.A.; Molla, H. Design, fabrication and performance test of a fixed bed batch type pyrolysis plant with scrap tire in Bangladesh. *J. Radiat. Res. Appl. Sci.* **2018**, *11*, 311–316. [CrossRef]
40. Diez, C.; Martinez, O.; Calvo, L.F.; Cara, J.; Morán, A. Pyrolysis of tires. Influence of the final temperature of the process on emissions and the calorific value of the products recovered. *Waste Manag.* **2004**, *24*, 463–469. [CrossRef] [PubMed]
41. Roy, C.; Chaala, A.; Darmstadt, H. The vacuum pyrolysis of used tires: End-uses for oil and carbon black products. *J. Anal. Appl. Pyrolysis* **1999**, *51*, 201–221. [CrossRef]

© 2020 by the authors. Licensee MDPI, Basel, Switzerland. This article is an open access article distributed under the terms and conditions of the Creative Commons Attribution (CC BY) license (http://creativecommons.org/licenses/by/4.0/).

Article

Estimating the Optimum Tilt Angles for South-Facing Surfaces in Palestine

Ramez Abdallah [1], Adel Juaidi [1,*], Salameh Abdel-Fattah [1] and Francisco Manzano-Agugliaro [2]

1 Mechanical Engineering Department, Faculty of Engineering & Information Technology, An-Najah National University, P.O. Box 7, 00970 Nablus, Palestine; ramezkhaldi@najah.edu (R.A.); salabdel@najah.edu (S.A.-F.)
2 Department of Engineering, CEIA3, University of Almeria, 04120 Almeria, Spain; fmanzano@ual.es
* Correspondence: adel@najah.edu

Received: 7 December 2019; Accepted: 27 January 2020; Published: 1 February 2020

Abstract: The optimum tilt angle of solar panels or collectors is crucial when determining parameters that affect the performance of those panels. A mathematical model is used for determining the optimum tilt angle and for calculating the solar radiation on a south-facing surface on a daily, monthly, seasonal, semi-annual, and annual basis. Photovoltaic Geographical Information System (PVGIS) and Photovoltaic Software (PVWatts) is developed by the NREL (US National Renewable Energy Laboratory) are also used to calculate the optimum monthly, seasonal, semi-annual, and annual tilt angles and to compare these results with the results obtained from the mathematical model. The results are very similar. PVGIS and PVWatts are used to estimate the solar radiation on south-facing surfaces with different tilt angles. A case study of a mono-crystalline module with 5 kWP of peak power is used to find out the amount of increased energy (gains) obtained by adjusting the Photovoltaic (PV) tilt angles based on yearly, semi-annual, seasonal, and monthly tilt angles. The results show that monthly adjustments of the solar panels in the main Palestinian cities can generate about 17% more solar energy than the case of solar panels fixed on a horizontal surface. Seasonal and semi-annual adjustments can generate about 15% more energy (i.e., it is worth changing the solar panels 12 times a year (monthly) or at least 2 times a year (semi-annually). The yearly optimum tilt angle for most Palestinian cities is about 29°, which yields an increase of about 10% energy gain compared to a solar panel fixed on a horizontal surface.

Keywords: optimal tilt angle; PV system; solar photovoltaic; solar radiation; PVGIS; PVWatts; Palestine

---

1. Introduction

Solar energy plays a crucial role in the future of sustainable energy as an environmentally friendly source of energy. Solar panel use has gained acceptance nowadays to become part of our modern-day life. The situation of energy in Palestine is different from that of surrounding countries due to several reasons: a scarcity of natural resources, the political conditions in Palestine are unstable, the difficult financial situation, and the high density of the population [1]. Palestine's natural resources and mineral wealth are scarce; this scarcity includes traditional sources of energy, such as oil and gas deposits. The high prices of hydrocarbon fuels in Palestinian cities are equivalent to those prices found in the most expensive cities in the world [1]. In addition, Palestine imports 100% of its fossil fuel and 87% of its electricity from other countries [1].

Palestine desperately needs all kinds of energy for growing its weak economy. There is no access to electricity for all Palestinian people for the whole day [1]. Palestinians strongly seek to overcome this dilemma by finding alternative sources of conventional fuel [2]. As in the case of the civilized world, they have found alternative sources of energy through which they hope to reduce their dependence on other countries. An example of this is the fuel crisis that the Gaza Strip experienced in mid-2013, which negatively affected the lives of people, especially in the health and transportation sectors; up to this time, Gaza Strip populations were living outside modern world era using primitive methods of mobility and lighting that disappeared a long time ago [2,3].

There is no doubt that the Palestinian tendency has shifted toward reliance on renewable sources of energy, which is consistent with the increasing global trends in the exploitation of alternative sources of energy. The trend in Palestine is concentrated in solar, biomass, wind, and geothermal energy [4]. The global trend is to find a clean, renewable, domestic sustainable energy source. This trend philosophy is increasing day by day, after significant damage caused by fossil fuels and their obvious risks to human health and the environment; toxic fumes and harmful gases from fossil fuel burning are a lingering health hazard endangering different forms of life on this planet. The most prominent environmental impact manifestations can be seen and felt right now, such as global warming, climate variability, ozone hole expansion, and acid rain, well as the frequent pollution of the seas by oil spills. The surge of oil prices is another driving force to find a replacement for conventional fossil fuels.

The Palestinians have succeeded to some extent in the exploitation of solar energy; they been using solar water heaters to heat water for domestic household uses and for limited industrial applications since the mid-seventies of the last century. The solar heater is a key component of every Palestinian home [5]. On the other hand, solar electricity generation remains limited due to research issues or limited activities of donors to help the population in disadvantaged areas. There are initial attempts at the exploitation of wind energy, as well as thermal energy. All these applications are still limited in the embryonic stage, with a promise of good potential [6,7].

1.1. Physical Characteristics of Palestine

Palestine, as it exists today, is formed from two separated land areas. The first is the West Bank with a total area of 5661 km^2, and the other is the Gaza Strip with a total area of 362 km^2 [8]. The West Bank is bordered to the north, west, and south by Israel and to the east by the Jordan River. It consists of eleven governorates: Jerusalem (East Jerusalem), Ramallah, Al-Bireh, Hebron, Nablus, Jenin, Tulkarm, Jericho, Tubas, Qalqiliya, Bethlehem, and Salfit. The Gaza Strip is located in the eastern Mediterranean Sea [8]. The Gaza Strip is bordered to the north and east by Israel, to the south by Egypt, and to the west by the Mediterranean Sea. It consists of five governorates: Rafah, Khan Yunis, North Gaza, Deir al Balah, and Gaza (as shown in the map in Figure 1).

Climate: Palestine has a Mediterranean climatic zone which has hot, long, dry summers and cool, short, wet winters [8]. However, the south of the Jordan Valley has a different climate zone and lies between the extreme desert conditions and dry steppe [8]. The climate within Palestine is affected by orographic effects, distance from the Mediterranean, altitude, and latitudinal position. Therefore, the climate of the West Bank can be divided into four main climatic zones: the Eastern Slopes, the Western Slopes, the Jordan Valley, and the Central Highlands. The climate of the West Bank, especially in the south, is affected by the vast nearby Arabian and the Negev deserts, especially during early summer and spring. Desert storms with hot winds full of dust and sand (locally known as Khamaseen winds) lead to increasing air temperatures and decreased humidity [8].

Temperature: The temperature of Palestine is comparatively high. The Jordan Valley and Jericho have the highest temperatures. The temperatures of the Jordan Valley are related inversely to altitude and decrease from south to north, with the highest temperature in the southern part at the Dead Sea. The Dead Sea is considered the lowest point on earth and is surrounded by a series of high mountains from both west and east, which create a natural greenhouse climate [8]. In summer, June, July, and August, the monthly average temperatures for the West Bank range from 20.8 °C to 30 °C. During winter months, December, January, and February, the monthly average temperatures in the West Bank range between 8.7 °C and 14.7 °C. The Gaza Strip lies between the semi-humid Mediterranean climate along the coast and the arid desert climate of the Sinai Peninsula. The monthly mean temperatures range from 13 °C in winter to 25 °C in summer [8].

Sunshine Duration: In Palestine, solar radiation varies from one city to another. June has the longest duration of sunlight, and December has the shortest duration of sunlight [8]. During the summer months, clear skies strengthen solar radiation, whereas during the winter, cloud cover reduces solar radiation [8].

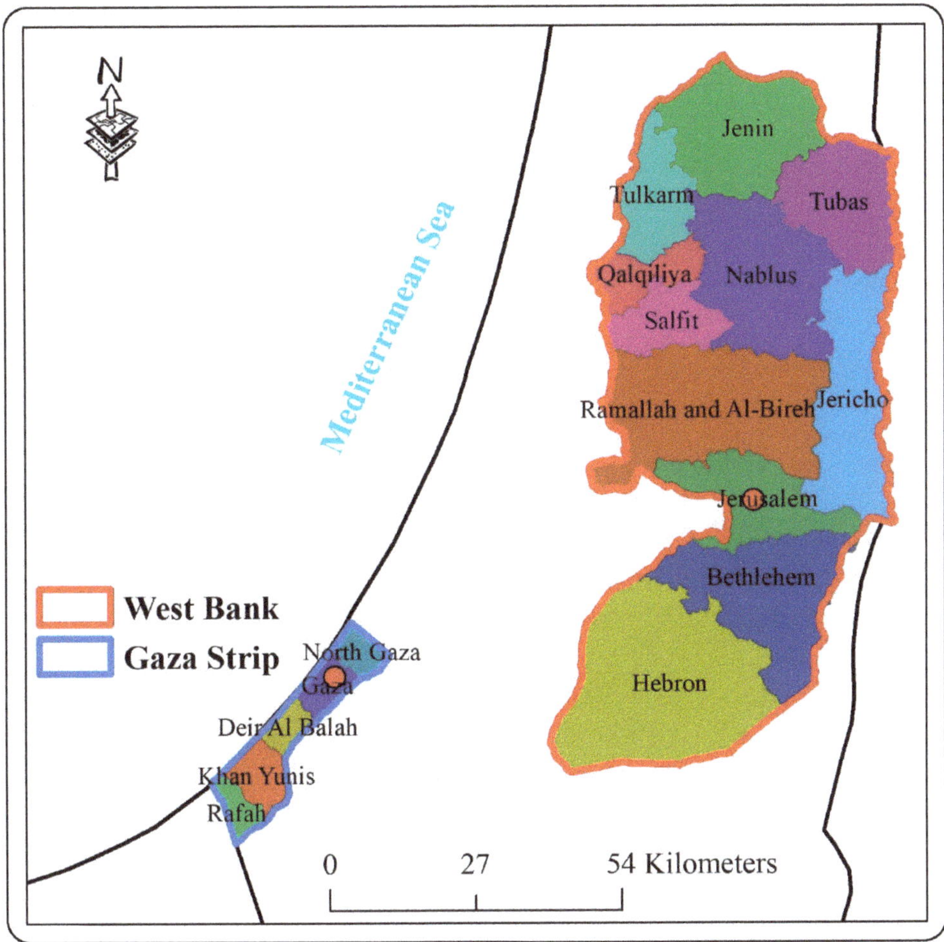

Figure 1. Map of the West Bank and Gaza Strip (Palestine).

1.2. The Situation of Electric Energy in Palestine and Net-Metering Regulations

The total electricity consumption for Palestine is 5137 GWh/year. About 87% is imported from the Israel Electric Corporation (IEC), while less than 8% is from the Palestine Electric Company. A small portion is imported from Jordan to supply Jericho in West Bank and from Egypt to supply Rafah in the Gaza Strip [1]. According to the Palestinian Central Bureau of Statistics (PCBS), more than 99.7% of the Palestinians in the West Bank and Gaza Strip are connected to the grid. For that reason, most of the solar PV in Palestine is an on-grid solar system [9].

The Palestinian Energy Authority (PEA) policy is to encourage the Palestinian people to utilize solar PV power to contribute to independency of Palestinian energy and support the Palestinian economy. To encourage solar PV energy, the PEA introduced the net-metering regulations.

The electricity generated by PV is supplied to the houses, where the grid serves as 'electricity storage'. In the case where solar PV produces more energy than that consumed by the houses, then the excess electrical energy is transferred to the grid. On the other hand, when the house electricity consumption is higher than the production, the house uses electricity from the grid. If the injection energy is more than the consumption during the month, 75% of the excess energy is credited to the next month. The credits are valid for 36 months. In times of higher consumption than production, the balance has to be paid by the consumer [10].

2. Methodology

Palestine is one of the countries enclosed between latitudes 40 degrees north and 40 degrees south (the so-called sun-belt). Palestine has more than 300 sunny days per year at an average brightness of 8 hours per day, increasing in summer and decreasing in winter. Previous studies show that the average daily solar radiation in Palestine reaches 5.4 kWh/m^2, which is equivalent to an annual production of 1950 kWh of energy, which has enabled Palestine to be one of the best areas using solar energy, and the possibility of investing in it is economically feasible [5].

When installing photovoltaic cells in an area, one must first calculate solar radiation in this region to optimize installation layouts to maximize the generated electricity, and to calculate solar radiation first, we must recognize the so-called solar angles.

Solar Angles

Declination angle (δ): is the angle between the equatorial plane and the line that connects between the earth's center and the sun's center. The declination angle varies seasonally between 23.45 and −23.45 and is calculated from the following relationship.

$$\delta = -23.45\, cos\left((n + 10.5)\frac{360}{365}\right), \tag{1}$$

n = the day number, such that n = 1 on the 1st of January.

Latitude angle (∅): is the angle between the equator and the line passing through the earth's center and a point on the earth's surface; northern latitudes are positive, and southern latitudes are negative.

Hour angle (h): is the angle between the plane containing the axis of the earth and the zenith and the other plane containing the axis of the earth and a point on the Earth's surface.

The altitude angle (α): is the angle between the horizontal and the rays of the sun. Therefore, the altitude angle for sunrise and sunset equals zero. The altitude angle (α) is calculated by [11]:

$$\sin\alpha = \sin\delta \sin\varnothing + \cos\delta \cos \mathrm{h} \cos\varnothing. \tag{2}$$

Note that the hour angles at sunrise and sunset (h_s) have the same magnitude and can be calculated by substituting α = 0 into the previous equation (Equation (2)) to obtained:

$$\mathrm{h_s} = \cos^{-1}[-\tan(\varnothing)\tan(\delta)]. \tag{3}$$

The tilt angle (β): is the angle between the horizontal and the tilted surface.

The total solar radiation is the sum of the direct, diffuse, and reflected solar radiation. Firstly, direct radiation depends on the atmosphere's thickness, the quantity of water vapor, and the level of pollution in the atmosphere. Secondly, diffuse solar radiation is defined as all the solar radiation that comes from the sky except for the part that comes directly from the sun. Therefore, it is difficult to calculate because it totally depends on highly variable cloudiness and atmospheric clarity. Finally, reflected radiation from buildings, cars, the ground, and streets must be calculated individually. For that reason, reflected radiation is usually neglected. In this paper, an "ASHRAE standard atmosphere" is assumed. This assumption will overestimate the solar radiation. On the other hand, this paper studies the optimum tilt angle rather than the available radiation. The total radiation obtained by the "ASHRAE standard atmosphere" assumption will be more than the actual available radiation, but the optimum tilt angles to achieve maximum radiation from any tilted surface will be the same. At the end of this study, the results will be compared to previously published papers.

In the Northern Hemisphere, it is considered that optimum orientation is due south [12,13]. Moreover, in 2013, Jafarkazemi and Saadabadi [14] found that the optimum orientation for Abu Dhabi was due south. Also, in 2010, Pavlović, et al. [15] estimated the optimum orientation for Serbia to be due south. Finally, in 2018, Omar and Mahmoud [16] confirmed that the optimum orientation for Palestine is due south. Based on the previous studies, the present mathematical model considers the titled surfaces to be due south.

The daily total extraterrestrial radiation on a tilted surface with an angle surface β and facing south is [9]:

$$I_d = \frac{24}{\pi} I_o\left[1 + 0.034\cos\left(\frac{2\pi n}{365}\right)\right] \times [\cos(\varnothing - \beta)\cos(\delta)\sin(h_s) + h_s\sin(\varnothing - \beta)\sin(\delta)]. \tag{4}$$

From Equation (4), for a particular day n to have an optimum tilt angle ($\beta_{op,d}$), the total radiation Id must be derived with respect to the tilt angle (β) and equalized with the resulting derivative set to zero, i.e., $\frac{dI_d}{d\beta} = 0.0$. to obtain:

$$\beta_{op,d} = \varnothing - tan^{-1}\left[\frac{hs}{sin(hs)}tan(\delta)\right]. \tag{5}$$

It could be concluded that adjusting the surface tilt angle every day is not practical, but it can be changed monthly or seasonally; the total radiation for a particular period can be obtained as:

$$I_p = \sum_{n=n_1}^{n=n_2} I_d\,, \tag{6}$$

where p is the particular period, n1 and n2 are the first and last days of a particular period, respectively. For obtaining the optimum tilt angle $\beta_{op,p}$ for a particular period (from n1 to n2), Ip must be derived with respect to the tilt angle (β) and equalized with the resulting derivative set to zero, i.e., $\frac{dI_p}{d\beta} = 0.0$ to obtain:

$$\beta_{op,p} = \varnothing - tan^{-1}\left[\frac{\sum_{n=n_1}^{n=n_2}\frac{24}{\pi}I_o\left[1 + 0.034cos\left(\frac{2\pi n}{365}\right)\right]sin(\delta)hs}{\sum_{n=n_1}^{n=n_2}\frac{24}{\pi}I_o\left[1 + 0.034cos\left(\frac{2\pi n}{365}\right)\right]cos(\delta)sin(hs)}\right]. \tag{7}$$

A MATLAB program (An-Najah National University Computer Labs.) was created to find the optimum tilt angle for any particular period using Equations (1)–(7). Using this program, the optimum tilt angle and total extraterrestrial radiation can be obtained for daily, monthly, seasonal, and yearly values. The major solar angles are shown in Figure 2a,b.

A comparison of the optimum tilt angle of the previous mathematical model was performed with the optimum tilt angle derived from the National Renewable Energy Laboratory (NREL) PVWatts program (NREL, 2017) [17] and PVGIS has been developed at the European Commission Joint Research Centre, at the JRC site in Ispra, Italy since 2001 [18]. The PVWatts Calculator is an online calculator that gives the amount of global radiation and the energy production of the PV using solar resource data for fixed locations throughout the world. The PVGIS gives the amount of global radiation and the energy production of the PV using the solar radiation data produced from satellite images. For the version of PVGIS used in this study, the satellite data used for the solar radiation estimates are from the METEOSAT satellites covering Europe, Africa, and most of Asia. Depending on satellite images, these images are captured every 15 minutes or 30 minutes [18]. Utilizing PVGIS, one image per hour was used; this means that PVGIS can be used for any location regardless of the distance to a particular meteorological station.

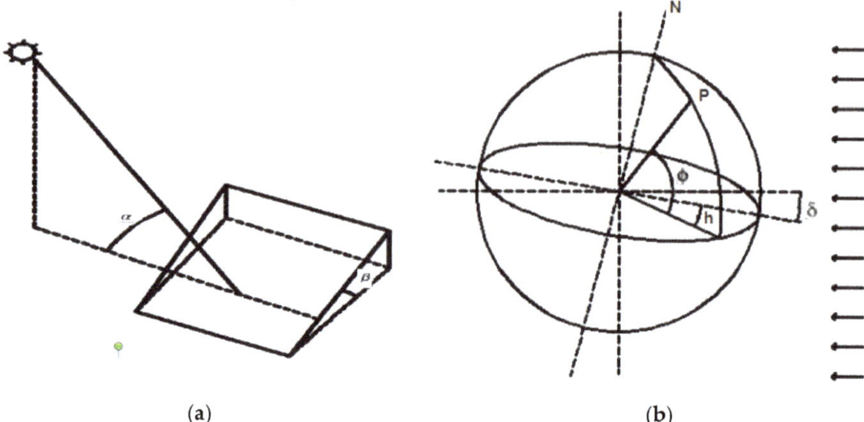

Figure 2. Major solar angles: (**a**) Tilt angle (β) and Altitude angle (α) (**b**)Declination angle (δ), Latitude angle (∅), and Hour angle (h).

PVWatts and PVGIS are used to calculate the solar radiation on a horizontal surface with several tilt angles. A case study of 5 kWP of peak power is used to demonstrate the importance of the optimum tilt angles.

3. Literature Review

With a constant orientation facing south, one of the most important factors that affects the performance of a solar panel or collector is its tilt angle since changing the tilt angle changes the amount of solar radiation received by the solar panel or collector.

The performance of solar equipment (i.e., solar collectors for water and air heating, power generation, photovoltaic systems, etc.), is closely related to the inclined angle from the surface that absorbs the light (i.e., with solar equipment placement plane) [19].

The maximum daily energy can be achieved using solar tracking systems. A mechanical device should be added to follow the direction of the sun on its daily sweep. These devices are expensive, need additional energy for operation, and are not always applicable [20,21]; therefore, it is often more practical to direct the surface at an optimum tilt angle, β. Moreover, according to the Palestinian Central Bureau of Statistics, there are no solar tracking projects in Palestine [9].

The optimum tilt angle must be changed daily, monthly, seasonally, semi-annually or annually to maximize the amount of solar radiation received by solar equipment [22].

Solar energy should be given more attention because it its environmentally friendly with no pollution and is a sustainable source of energy. The amount of solar radiation changes with the latitude of the location, the day number through the years, and time of a day. There are still lots of errors during the installation of photovoltaic (PV) panels; significant errors are made by companies and personnel. These errors are wrong PV cell selection, wrong installation spaces and orientation of the PV panels, and also wrong tilt angles [23]. Jiménez-Torres, et al. stated that information of the available solar resource in a particular location on a surface in any position (inclination and orientation) is essential [24]. Mousavi, et al. mentioned that the efficiency of a PV panel can be optimized if the panel is positioned in such a way that it receives the maximum amount of incident solar radiation [25]. Table 1 provides the optimal tilt angle results in different countries of the world.

Table 1. Optimal tilt angle results in different regions of the world.

Location	Tilt Angle (Optimum in Degree)	Latitude	Reference
Damascus	30.56°	33.5°	[26]
Northern Jordan	30° Average 10° Summer 50° Winter	32.5°	[27]
Austria	30°	47.5°	[28]
Germany	45°	51.2°	[28]
Tehran	35.7°	35.7°	[29]
Isfahan	32°	36.5°	[29]
Shiraz	29.4°	29.6°	[29]
Mashhad	36.2°	36.3°	[29]
Tabriz	38°	38.1°	[29]
Pristina	8.9° Summer 25.7° Spring 50.9° Autumn 62.1° Winter 34.7° Average	42.7°	[19]
Cyprus	20° Summer 50° Winter	35.1°	[23]
Montreal	37°	45.5°	[30]
Bordeaux	33°	44.8°	[30]
Cologne	32°	50.9°	[30]
Hong Kong	20°	22.4°	[30]
Rajko	24°	42.1°	[30]
Zahedan	26.70°	29.5°	[30]
Birjand	29.93°	32.9°	[31]
Shraz	25.88°	52.9°	[31]
Tabas	30.16°	33.6°	[31]
Yazd	29.05°	31.9°	[31]
Kerman	23.95°	36.7°	[31]
Assiut	27°	27.2°	[32]
Valencia	31°	39°	[33]
Darussalam	3.3°	4.5°	[34]
Beijing	39.2°	39.9°	[35]
Kunming	27.9°	24.9°	[35]
Shanghai	28°	31.2°	[35]
Guangzhou	22°	23.1°	[35]
Chengdu	23°	30.6°	[35]
Xi'an	30.1°	34.3°	[35]
Yinchuang	38.3°	38.5°	[35]
Shenyang	40.3°	41.8°	[35]
Izmir	0° June 61° December	38.4°	[36]

Table 1. *Cont.*

Location	Tilt Angle (Optimum in Degree)	Latitude	Reference
Gaza Strip	32.1°	31.5°	[37]
Damascus	33.7°	33.5°	[37]
Beirut	33.8°	33.9°	[37]
Tunis	35.2°	33.9°	[37]
Seville	36.6°	37.4°	[37]
Milan	41.8°	45.5°	[37]
Mu'tah	28.5°	31.7°	[11]
Sanya	−18.1° June 49.9° December	−11.5°	[38]
Shanghai	−7.6° June 61.4° December	31.2°	[38]
Zhengzhou	5.5° June 64.3° December	34.7°	[38]
Harbin	12.6° June 73.7° December	45.8°	[38]
Mohe	16.6° June 80.0° December	52.9°	[38]
Lhasa	−8.9° June 59.9° December	29.7°	[38]

4. Results and Discussion

Since more than 99.7% of the Palestinian people in the West Bank and Gaza Strip are connected to the grid and the majority of the PVs in Palestine are an on-grid solar system in which the grid serves as electricity storage [9], the optimum tilt angle is important to produce the maximum energy generation during the month regardless of the peak load time during the day. The advantages of on-grid PV solar systems include the benefits of getting rid of the difficulties and energy losses associated with energy storage and the peak load consumption vs. the maximum power generation from the PV system.

Considering that most Palestinian communities are connected to the grid and the electricity tariff is constant during the day, the maximum energy production during the year is a vital factor for PV applications. In the present study, the optimum tilt angle is calculated to achieve this maximum energy production.

The daily, monthly, seasonal, semi-annual, and annual optimum tilt angles are calculated using the mathematical model. Also, a monthly, seasonal, semi-annual, and annual optimum tilt angle is calculated for two Palestinian cities, Jerusalem in the West Bank and Gaza city in the Gaza Strip, using PVWatts and PVGIS.

According to PEA regulations, 5 kWP is the maximum allowed peak power PV for houses connected to the grid in Palestine. The annual energy consumption for most residential houses does not exceed 5 kWP PV. Therefore, most houses implement 5 kWP. Moreover, there are no specific standards or specifications for selecting PV modules, but all the PV modules in Palestine are a polycrystalline or monocrystalline type. For that reason, in the present study, 5 kWP will be taken as a case study.

4.1. Part One: Mathematical Model Results

4.1.1. Daily Optimum Tilt Angle

Equations (1), (3), (4), and (5) are used in the MATLAB program to calculate the daily optimum tilt angle $\beta_{opt,d}$ and the daily total extraterrestrial radiation for different latitudes and the main Palestinian cities. The results obtained from the MATLAB program are tabulated for recommended average days of months in Table 2 for different latitude angles and in Table 3 for the main Palestinian cities and are plotted in Figure 3 for different latitude angles as well as in Figure 4 for Jerusalem and Gaza city. The results show that the daily optimum tilt angle $\beta_{opt,d}$ increases with the increasing latitude of the selected location, and it changes with the number of day which takes the form of a triangular cosine function. Moreover, the optimum daily angle $\beta_{opt,d}$ for 21 March is slightly more than the latitude of the location whereas $\beta_{opt,d}$ for 22 March is slightly less than the latitude of the location. On the other hand, the optimum daily angle $\beta_{opt,d}$ for 20 September is slightly less than the latitude of the location whereas $\beta_{opt,d}$ for 21 September is slightly more than the latitude of the location. The same results were obtained for the main Palestinian cities. Daily optimum tilt angles $\beta_{opt,d}$ for the Palestinian cities are plotted in Figure 3. It can be concluded that negative values for $\beta_{opt,d}$ for the main Palestinian cities were obtained for the time period from n = 134 to 210.

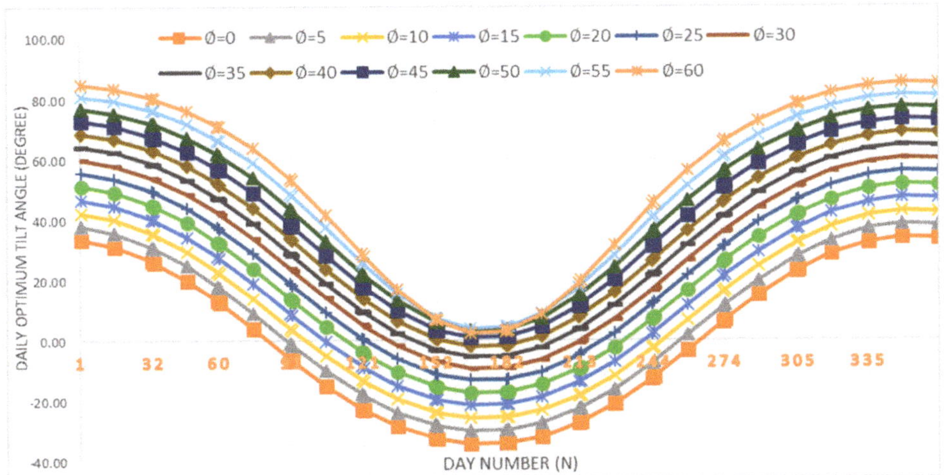

Figure 3. The daily optimum tilt angle $\beta_{opt,d}$ for different latitude angles.

4.1.2. Monthly Optimum Tilt Angle

Equations (1), (3), (4), (6), and (7) are used in the MATLAB program to calculate the monthly optimum tilt angle ($\beta_{opt,m}$) and the monthly total extraterrestrial radiation (I_m) for different latitude angles and for the main Palestinian cities. The results obtained from the MATLAB program are tabulated in Table 4 for different latitude angles and are plotted in Figure 5. It important to mention that when the optimum monthly tilt angle $\beta_{opt,m}$ is negative for a particular month, then the monthly total extraterrestrial radiation for that month is recalculated by considering that the optimum monthly tilt angle $\beta_{opt,m}$ is equal to zero. The reason for that is to determine if it is worth designing a PV or solar collector with a negative tilt angle. The negative sign means that the solar panel or collector is north facing, while the positive sign means that the solar panel or collector is south facing. From Figure 4, it can be noted that the variation of the monthly optimum tilt angle can be described by a triangular cosine function. The results obtained for the main Palestinian cities are tabulated in Table 5. The $\beta_{opt,m}$ for Jerusalem is plotted in Figure 6. From Table 5 it can be concluded that $\beta_{opt,m}$ increases with increasing latitude angle. Moreover, for latitude 0, $\beta_{opt,m}$ has a negative value from April to September, while for latitudes 5, 10, and 15, $\beta_{opt,m}$ has a negative value from April to August. Also, for latitude 20, $\beta_{opt,m}$ has a negative value for May to August. Additionally, for latitudes 25 and 30, $\beta_{opt,m}$ has a negative value for May to July. For latitude 35, $\beta_{opt,m}$ has a negative value for June and July, and finally, for latitude 40, $\beta_{opt,m}$ has a negative value for only the month of June. The optimum tilt angle in March is slightly higher than the latitude but is slightly lower than the latitude in September.

Table 2. The daily optimum tilt angle $\beta_{opt,d}$ for recommended average days of months for different latitude angles.

Date	Day Number (n)	Latitude (°)												
		0	5	10	15	20	25	30	35	40	45	50	55	60
17 Jan.	17	30.92	35.39	39.88	44.37	48.86	53.34	57.80	62.23	66.61	70.94	75.19	79.34	83.33
16 Febr.	47	19.74	24.51	29.29	34.06	38.83	43.59	48.34	53.08	57.79	62.46	67.09	71.66	76.15
16 Mar.	75	3.64	8.63	13.62	18.61	23.60	28.59	33.58	38.57	43.56	48.55	53.53	58.51	63.48
15 April	105	−14.74	−9.87	−5.01	−0.16	4.69	9.52	14.33	19.11	23.85	28.54	33.15	37.64	41.92
15 May	135	−28.21	−23.67	−19.16	−14.68	−10.26	−5.90	−1.64	2.49	6.44	10.12	13.36	15.87	16.98
11 June	162	−33.83	−29.47	−25.15	−20.89	−16.71	−12.63	−8.69	−4.96	−1.51	1.50	3.81	4.86	3.32
17 July	198	−31.27	−26.83	−22.43	−18.07	−13.77	−9.56	−5.47	−1.54	2.15	5.47	8.23	10.00	9.80
16 Aug.	228	−20.48	−15.73	−10.99	−6.27	−1.58	3.09	7.71	12.27	16.74	21.10	25.27	29.15	32.50
15 Sept.	258	−3.32	1.67	6.66	11.66	16.65	21.64	26.63	31.62	36.61	41.60	46.58	51.56	56.54
15 Oct.	288	15.02	19.88	24.75	29.61	34.47	39.33	44.18	49.01	53.83	58.63	63.40	68.13	72.80
14 Nov.	318	28.37	32.92	37.48	42.04	46.60	51.15	55.68	60.19	64.65	69.06	73.41	77.65	81.76
10 Dec.	344	33.78	38.17	42.57	46.98	51.39	55.78	60.16	64.50	68.79	73.03	77.17	81.20	85.06

Table 3. The daily optimum tilt angle $\beta_{opt,d}$ for recommended average days of months for the main Palestinian cities.

Date	City									
	Day Number (n)	Jenin	Tulkarm	Nablus	Ramallh	Jerusalem	Hebron	Jerico	Gaza	Rafah
17 Jan.	17	59.98	59.85	59.77	59.49	59.38	59.46	59.16	59.16	58.95
16 Febr.	47	50.68	50.53	50.45	50.14	50.03	50.12	49.79	49.79	49.57
16 Mar.	75	36.04	35.89	35.80	35.48	35.36	35.45	35.11	35.11	34.87
15 April	105	16.68	16.54	16.45	16.15	16.03	16.12	15.79	15.79	15.56
15 May	135	0.41	0.29	0.21	−0.05	−0.15	−0.08	−0.36	−0.36	−0.56
11 June	162	−6.82	−6.94	−7.00	−7.25	−7.34	−7.27	−7.52	−7.52	−7.71
17 July	198	−3.51	−3.63	−3.70	−3.95	−4.05	−3.98	−4.25	−4.25	−4.44
16 Aug.	228	9.96	9.82	9.74	9.45	9.34	9.42	9.11	9.11	8.89
15 Sept.	258	29.09	28.94	28.85	28.53	28.41	28.50	28.16	28.16	27.92
15 Oct.	288	46.56	46.41	46.33	46.02	45.90	45.99	45.66	45.66	45.43
14 Nov.	318	57.90	57.77	57.69	57.40	57.29	57.37	57.06	57.06	56.85
10 Dec.	344	62.30	62.17	62.09	61.81	61.71	61.79	61.49	61.49	61.28

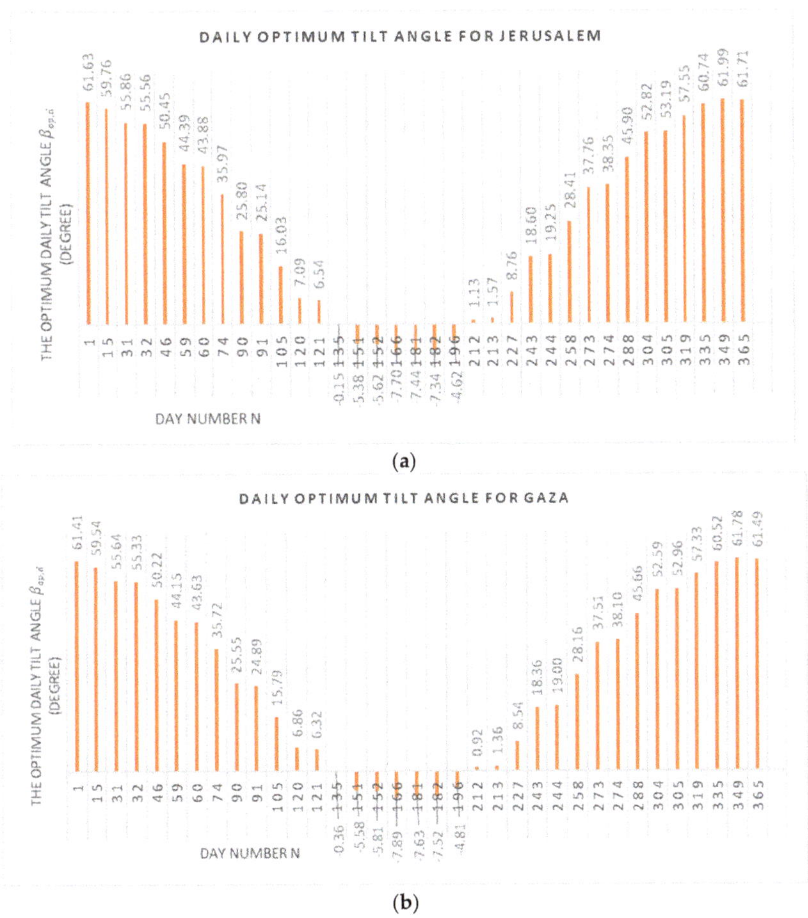

Figure 4. The daily optimum tilt angle $\beta_{opt,d}$ for Jerusalem (**a**) and Gaza City (**b**).

From Table 5, it can be concluded that for the main cities of Palestine, the monthly optimum tilt angle is between -7.0 (in June) to 62.0 (in December). Moreover, it can be concluded that for the main cities of Palestine, the optimum tilt angle for March is slightly higher than the latitude of the city ($\beta_{opt,m}$ = 35), whereas for September it is slightly less than the latitude of the city ($\beta_{opt,m}$ = 29°). The yearly extraterrestrial radiation is calculated with $\beta = 0$ and $\beta = \beta_{opt}$ on a daily, monthly, seasonal, semi-annual, and annual basis. The tilt factor is defined as the ratio between the radiation on a tilted surface and the radiation on a horizontal surface in the same period. The tilt factor is calculated on a daily, monthly, seasonal, semi-annual, and annual basis, and the results are shown in Table 6. Moreover, the tilt factors for daily, monthly, seasonal, semi-annual, and annual optimum tilt angles for the main Palestinian cities are presented in Table 7. It can be concluded from Tables 6 and 7 that the tilt factor on a daily basis is almost the same as that on a monthly basis. This means that it worth changing the tilt angle monthly and there is no need to change it daily.

Table 4. The monthly optimum tilt angle $\beta_{opt,m}$ and extraterrestrial radiation I_m in kWh/m^2 for different latitudes.

Month		Latitude												
		0	5	10	15	20	25	30	35	40	45	50	55	60
January	$\beta_{opt,m}$	30.83	35.31	39.79	44.28	48.77	53.25	57.70	62.13	66.51	70.84	75.08	79.21	83.17
	I_m	363.69	361.52	359.03	356.17	352.82	348.86	344.07	338.15	330.63	320.73	307.07	287.01	254.55
February	$\beta_{opt,m}$	20.30	25.05	29.80	34.55	39.30	44.04	48.76	53.47	58.15	62.79	67.38	71.90	76.32
	I_m	310.38	309.81	309.11	308.27	307.24	305.99	304.45	302.51	300.01	296.71	292.17	285.62	275.49
March	$\beta_{opt,m}$	3.62	8.59	13.57	18.54	23.51	28.48	33.44	38.41	43.36	48.32	53.26	58.19	63.10
	I_m	326.49	326.48	326.45	326.40	326.33	326.25	326.13	325.98	325.77	325.48	325.08	324.50	323.59
April	$\beta_{opt,m}$	−14.83	−9.98	−5.14	−0.30	4.52	9.33	14.11	18.87	23.57	28.22	32.78	37.19	41.35
	I_m	316.49	316.67	316.79	316.83	316.79	316.66	316.41	316.01	315.38	314.44	312.99	310.74	307.03
I_m (negative values of $\beta_{opt,m}$ are considered zero)		305.95	311.88	315.51	316.82	316.79	316.66	316.41	316.01	315.38	314.44	312.99	310.74	307.03
May	$\beta_{opt,m}$	−28.21	−23.68	−19.18	−14.71	−10.29	−5.95	−1.70	2.42	6.35	10.00	13.21	15.66	16.63
	I_m	339.30	340.66	341.82	342.78	343.56	344.11	344.41	344.37	343.87	342.69	340.47	336.54	329.67
I_m (negative values of $\beta_{opt,m}$ are considered zero)		299.00	311.97	322.85	331.55	338.03	342.26	344.26	344.37	343.87	342.69	340.47	336.54	329.67
June	$\beta_{opt,m}$	−33.80	−29.45	−25.13	−20.87	−16.69	−12.61	−8.67	−4.93	−1.48	1.54	3.84	4.90	3.36
	I_m	335.61	337.94	340.02	341.89	343.54	344.98	346.18	347.08	347.59	347.56	346.75	344.86	341.86
I_m (negative values of $\beta_{opt,m}$ are considered zero)		278.87	294.28	307.83	319.46	329.08	336.67	342.22	345.79	347.47	347.56	346.75	344.86	341.86
July	$\beta_{opt,m}$	−31.17	−26.72	−22.32	−17.96	−13.66	−9.45	−5.36	−1.43	2.26	5.58	8.34	10.10	9.85
	I_m	341.46	343.31	344.93	346.34	347.54	348.52	349.23	349.61	349.54	348.82	347.13	343.89	338.27
I_m (negative values of $\beta_{opt,m}$ are considered zero)		292.18	306.63	319.09	329.47	337.71	343.79	347.71	349.50	349.54	348.82	347.13	343.89	338.27
August	$\beta_{opt,m}$	−20.23	−15.49	−10.76	−6.05	−1.36	3.29	7.90	12.45	16.91	21.24	25.38	29.21	32.49
	I_m	327.47	327.95	328.31	328.55	328.67	328.63	328.41	327.94	327.13	325.81	323.70	320.29	314.53
I_m (negative values of $\beta_{opt,m}$ are considered zero)		307.27	316.04	322.54	326.72	328.58	328.63	328.41	327.94	327.13	325.81	323.70	320.29	314.53
September	$\beta_{opt,m}$	−2.95	2.03	7.00	11.98	16.95	21.93	26.89	31.86	36.82	41.77	46.72	51.65	56.56
	I_m	310.24	310.24	310.23	310.21	310.17	310.11	310.04	309.93	309.79	309.58	309.29	308.85	308.16
I_m (negative values of $\beta_{opt,m}$ are considered zero)		309.83	310.24	310.23	310.21	310.17	310.11	310.04	309.93	309.79	309.58	309.29	308.85	308.16
October	$\beta_{opt,m}$	15.41	20.26	25.10	29.94	34.78	39.61	44.43	49.24	54.03	58.79	63.52	68.19	72.79
	I_m	333.11	332.82	332.44	331.96	331.36	330.62	329.70	328.52	326.98	324.94	322.12	318.04	311.76
November	$\beta_{opt,m}$	28.56	33.10	37.65	42.20	46.75	51.29	55.81	60.30	64.75	69.14	73.46	77.68	81.74
	I_m	344.87	343.21	341.30	339.07	336.46	333.34	329.56	324.87	318.91	311.06	300.26	284.49	259.38
December	$\beta_{opt,m}$	33.83	38.21	42.61	47.02	51.43	55.82	60.19	64.53	68.83	73.06	77.20	81.22	85.08
		370.09	367.20	363.94	360.22	355.91	350.83	344.73	337.21	327.66	315.07	297.63	271.71	228.61
	Total	4019.18	4017.79	4014.36	4008.69	4000.40	3988.92	3973.31	3952.18	3923.26	3882.89	3824.65	3736.52	3592.91
Total I_m (negative values of $\beta_{opt,m}$ are considered zero)		3841.70	3892.07	3930.32	3956.31	3970.48	3974.02	3967.68	3950.78	3923.15	3882.89	3824.65	3736.52	3592.91

Table 5. The monthly optimum tilt angle $\beta_{opt,m}$ and extraterrestrial radiation I_m in kWh/m^2 for the main Palestinian cities.

Month		City								
		Jenin	**Tulkarm**	**Nablus**	**Ramallah**	**Jerusalem**	**Jericho**	**Hebron**	**Ghaza**	**Rafah**
January	$\beta_{opt,m}$	59.89	59.75	59.67	59.39	59.28	59.36	59.06	59.06	58.85
	I_m	341.32	341.50	341.60	341.97	342.11	342.01	342.40	342.40	342.67
February	$\beta_{opt,m}$	51.08	50.94	50.85	50.55	50.44	50.52	50.20	50.20	49.98
	I_m	303.55	303.61	303.65	303.77	303.81	303.78	303.90	303.90	303.99
March	$\beta_{opt,m}$	35.89	35.74	35.65	35.33	35.21	35.30	34.96	34.96	34.72
	I_m	314.66	314.65	314.66	314.64	314.63	314.64	314.62	314.62	314.60
April	$\beta_{opt,m}$	16.46	16.31	16.23	15.92	15.81	15.89	15.57	15.57	15.34
	I_m	316.24	316.25	316.25	316.28	316.29	316.28	316.31	316.31	316.32
May	$\beta_{opt,m}$	0.35	0.23	0.15	−0.11	−0.21	−0.14	−0.42	−0.42	−0.62
	I_m	344.44	344.44	344.44	344.44	344.44	344.44	344.44	344.44	344.44
I_m (negative values of $\beta_{opt,m}$ are considered zero)		344.44	344.44	344.44	344.44	344.44	344.44	344.43	344.43	344.42
June	$\beta_{opt,m}$	−6.80	−6.91	−6.98	−7.22	−7.31	−7.24	−7.50	−7.50	−7.68
	I_m	346.66	346.63	346.62	346.56	346.53	346.55	346.49	346.49	346.44
I_m (negative values of $\beta_{opt,m}$ are considered zero)		344.22	344.11	344.05	343.81	343.72	343.79	343.52	343.52	343.33
July	$\beta_{opt,m}$	−3.40	−3.52	−3.59	−3.84	−3.94	−3.87	−4.14	−4.14	−4.33
	I_m	349.47	349.45	349.45	349.42	349.41	349.42	349.39	349.39	349.37
I_m (negative values of $\beta_{opt,m}$ are considered zero)		348.85	348.80	348.76	348.63	348.58	348.62	348.48	348.48	348.37
August	$\beta_{opt,m}$	10.15	10.01	9.93	9.64	9.53	9.61	9.30	9.30	9.08
	I_m	328.22	328.23	328.24	328.27	328.28	328.27	328.30	328.30	328.32
September	$\beta_{opt,m}$	29.34	29.19	29.10	28.78	28.66	28.75	28.41	28.41	28.18
I_m		320.26	320.26	320.27	320.27	320.27	320.27	320.28	320.28	320.28
October	$\beta_{opt,m}$	46.80	46.66	46.57	46.26	46.15	46.23	45.91	45.91	45.68
	I_m	329.15	329.19	329.21	329.28	329.31	329.29	329.37	329.37	329.42
November	$\beta_{opt,m}$	58.02	57.89	57.80	57.52	57.41	57.49	57.18	57.18	56.97
	I_m	327.39	327.52	327.61	327.90	328.01	327.93	328.23	328.23	328.45
December	$\beta_{opt,m}$	62.33	62.20	62.13	61.85	61.74	61.82	61.53	61.53	61.32
	I_m	341.24	341.46	341.60	342.07	342.24	342.11	342.60	342.60	342.95
	Total	3962.59	3963.21	3963.58	3964.86	3965.34	3964.99	3966.32	3966.32	3967.24
Total (negative values of $\beta_{opt,m}$ are considered zero)		3959.54	3960.03	3960.33	3961.33	3961.69	3961.43	3962.44	3962.44	3963.12

Table 6. The tilt factor for the daily, monthly, seasonal, semi-annual, and annual optimum tilt angles for different latitudes.

		Latitude												
		0	5	10	15	20	25	30	35	40	45	50	55	60
$\beta = 0$	I	3653.77	3637.62	3596.30	3530.22	3440.05	3326.73	3191.50	3035.91	2861.94	2672.05	2469.50	2258.81	2047.02
$\beta_{opt,d}$	I	4027.36	4025.97	4022.56	4016.93	4008.71	3997.32	3981.83	3960.87	3932.20	3892.18	3834.50	3747.30	3605.51
	F	1.1022	1.1068	1.1185	1.1379	1.1653	1.2016	1.2476	1.3047	1.3740	1.4566	1.5527	1.6590	1.7613
$\beta_{opt,m}$	I	4019.18	4017.79	4014.36	4008.69	4000.40	3988.92	3973.31	3952.18	3923.26	3882.89	3824.65	3736.52	3592.91
	F	1.1000	1.1045	1.1162	1.1355	1.1629	1.1990	1.2450	1.3018	1.3708	1.4531	1.5488	1.6542	1.7552
$\beta_{opt,s}$	I	3958.60	3957.14	3953.53	3947.57	3938.80	3926.72	3910.22	3887.86	3857.20	3814.27	3752.10	3657.37	3501.03
	F	1.0834	1.0878	1.0993	1.1182	1.1450	1.1804	1.2252	1.2806	1.3478	1.4275	1.5194	1.6192	1.7103
$\beta_{opt,sa}$	I	3952.82	3951.37	3947.71	3941.62	3932.68	3920.26	3903.33	3880.34	3848.79	3804.52	3740.23	3641.86	3478.27
	F	1.0818	1.0863	1.0977	1.1165	1.1432	1.1784	1.2230	1.2781	1.3448	1.4238	1.5146	1.6123	1.6992
$\beta_{opt,y}$	I	3653.99	3652.54	3648.73	3642.33	3632.91	3619.77	3601.84	3577.44	3543.91	3496.82	3428.34	3323.42	3148.68
	F	1.0001	1.0041	1.0146	1.0318	1.0561	1.0881	1.1286	1.1784	1.2383	1.3087	1.3883	1.4713	1.5382

I: The yearly extraterrestrial radiation in kWh/m^2; F: The tilt factor.

Table 7. The tilt factor for the daily, monthly, seasonal, semi-annual, and yearly optimum tilt angles for the main Palestinian cities.

		City								
		Jenin	Tulkarm	Nablus	Ramallah	Jerusalem	Jericho	Hebron	Gaza	Rafah
$\beta = 0$	I	3117.38	3122.04	3124.83	3134.68	3138.35	3135.60	3145.97	3145.97	3153.24
$\beta_{opt,d}$	I	3972.32	3972.94	3973.31	3974.61	3975.09	3974.73	3976.08	3976.08	3977.02
	F	1.2742	1.2725	1.2715	1.2679	1.2666	1.2676	1.2639	1.2639	1.2613
$\beta_{opt,m}$	I	3962.59	3963.21	3963.58	3964.86	3965.34	3964.99	3966.32	3966.32	3967.24
	F	1.2711	1.2694	1.2684	1.2648	1.2635	1.2645	1.2608	1.2608	1.2581
$\beta_{opt,s}$	I	3900.08	3900.74	3901.14	3902.22	3903.04	3902.66	3904.09	3904.09	3905.09
	F	1.2511	1.2494	1.2484	1.2449	1.2437	1.2446	1.2410	1.2410	1.2384
$\beta_{opt,sa}$	I	3892.90	3893.59	3893.99	3895.42	3895.95	3895.55	3897.04	3897.04	3898.06
	F	1.2488	1.2471	1.2461	1.2427	1.2414	1.2424	1.2387	1.2387	1.2362
$\beta_{opt,y}$	I	3590.777454	3591.50	3591.93	3593.44	3594.00	3593.58	3595.16	3595.16	3596.25
	F	1.1519	1.1504	1.1495	1.1464	1.1452	1.1461	1.1428	1.1428	1.1405

I: The yearly extraterrestrial radiation in kWh/m^2; F: The tilt factor.

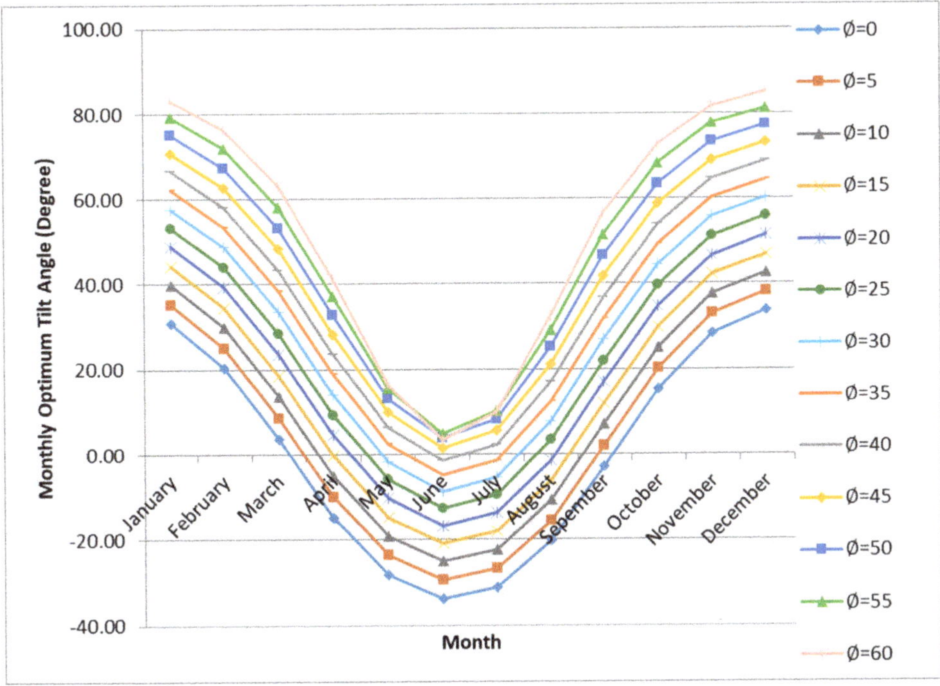

Figure 5. The monthly optimum tilt angle $\beta_{opt,m}$ for different latitudes.

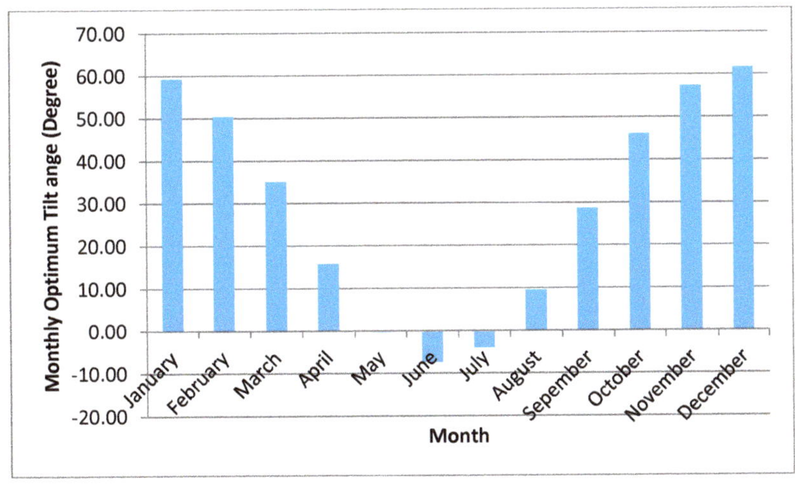

Figure 6. The monthly optimum tilt angle $\beta_{opt,m}$ for Jerusalem, Palestine.

4.1.3. Seasonal Optimum Tilt Angle

It is more practical to change the tilt angle quarterly rather than daily or monthly. Therefore, the tilt angle is changed once every three months. The Palestinian climate can be divided into four seasons: winter (December to February), spring (March to May), summer (June to August), and autumn (September to November). The seasonal optimum tilt angle and the seasonal total extraterrestrial radiation for different latitudes and for the main Palestinian cities are calculated. The seasonal variation of optimum angle $\beta_{opt,s}$, and the seasonal total extraterrestrial radiation Is for different latitude angles and for the main Palestinian cities are listed in Tables 8 and 9, respectively. Moreover, the seasonal variation of optimum angle $\beta_{opt,s}$ for different latitude angles and for Jerusalem are plotted in Figures 7 and 8, respectively. From Table 7, it can be shown that for most of the Palestinian cities, the solar radiation on a surface tilted by $\beta_{opt,s}$ is 25% more than the radiation falling on a horizontal surface ($\beta = 0$). Moreover, there is a decrease of less than 2% in the solar radiation on a surface tilted by the seasonal optimum tilt angle compared to a surface tilted by monthly optimum tilt angles for all the main Palestinian cities.

From Table 9, it could be concluded that the minimum and maximum solar radiation at a seasonally optimum tilt angle occurs in autumn and summer, respectively. For Jerusalem, the minimum and maximum radiation values are (947.13 kW h/m^2) and (1016.15 kW h/m^2), respectively.

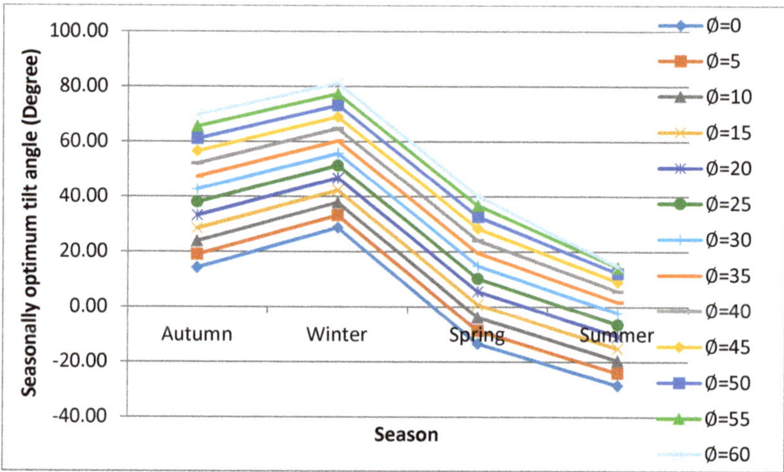

Figure 7. The seasonally optimum tilt angle $\beta_{opt,s}$ for different latitudes.

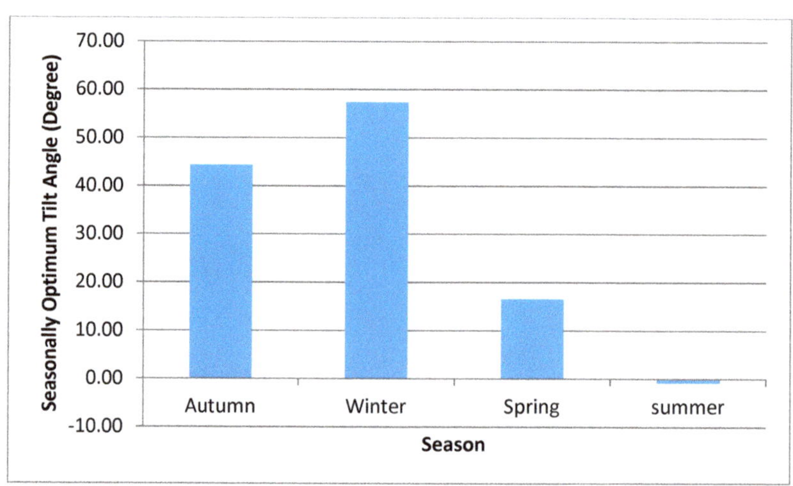

Figure 8. The seasonally optimum tilt angle $\beta_{opt,s}$ for Jerusalem, Palestine.

Table 8. The seasonally optimum tilt angles $\beta_{opt,s}$ and extraterrestrial radiation I_s in kWh/m^2 for different latitudes.

Season		Latitude												
		0	5	10	15	20	25	30	35	40	45	50	55	60
Autumn	$\beta_{opt,s}$	14.27	19.02	23.78	28.53	33.27	38.00	42.72	47.40	52.05	56.65	61.17	65.56	69.76
	I_s	963.57	962.33	960.72	958.67	956.11	952.89	948.83	943.63	936.81	927.61	914.72	895.64	865.08
Winter	$\beta_{opt,s}$	28.77	33.28	37.80	42.32	46.84	51.35	55.83	60.28	64.69	69.03	73.28	77.39	81.26
	I_s	1039.09	1033.75	1027.58	1020.41	1011.95	1001.94	989.75	974.63	955.33	929.81	894.47	842.27	756.98
Spring	$\beta_{opt,s}$	−13.34	−8.59	−3.84	0.90	5.62	10.31	14.96	19.56	24.09	28.50	32.76	36.74	40.22
	I_s	956.61	957.39	957.84	957.95	957.68	956.95	955.64	953.56	950.40	945.63	938.33	926.76	907.23
Is (negative values of $\beta_{opt,s}$ are considered zero)		930.78	946.66	955.70	957.95	957.68	956.95	955.64	953.56	950.40	945.63	938.33	926.76	907.23
Summer	$\beta_{opt,s}$	−28.49	−23.99	−19.52	−15.10	−10.73	−6.43	−2.24	1.81	5.65	9.18	12.20	14.35	14.72
	I_s	999.32	1003.66	1007.39	1010.53	1013.06	1014.93	1016.00	1016.04	1014.66	1011.22	1004.58	992.69	971.75
Is (negative values of $\beta_{opt,s}$ are considered zero)		878.32	916.95	949.46	975.65	995.37	1008.55	1015.22	1016.04	1014.66	1011.22	1004.58	992.69	971.75
Total		3958.60	3957.14	3953.53	3947.57	3938.80	3926.72	3910.22	3887.86	3857.20	3814.27	3752.10	3657.37	3501.03
Total (negative values of $\beta_{opt,s}$ are considered zero)		3811.76	3859.69	3893.46	3912.69	3921.11	3920.33	3909.45	3887.86	3857.20	3814.27	3752.10	3657.37	3501.03

Table 9. The seasonally optimum tilt angles $\beta_{opt,s}$ and extraterrestrial radiation I_s in kWh/m^2 for the main Palestinian cities.

Season		City								
		Jenin	Tulkarm	Nablus	Ramallah	Jerusalem	Jericho	Hebron	Gaza	Rafah
Autumn	$\beta_{opt,s}$	45.03	44.88	44.80	44.50	44.39	44.47	44.15	44.15	43.93
	I_s	946.44	946.59	946.68	947.01	947.13	947.04	947.38	947.38	947.61
Winter	$\beta_{opt,s}$	58.03	57.89	57.81	57.53	57.42	57.50	57.20	57.20	56.98
	I_s	982.74	983.19	983.46	984.10	984.76	984.50	985.48	985.48	986.17
Spring	$\beta_{opt,s}$	17.23	17.09	17.01	16.71	16.60	16.69	16.37	16.37	16.15
	I_s	954.73	954.79	954.83	954.96	955.00	954.97	955.10	955.10	955.19
summer	$\beta_{opt,s}$	−0.22	−0.34	−0.42	−0.68	−0.78	−0.70	−0.98	−0.98	−1.18
	I_s	1016.17	1016.16	1016.16	1016.15	1016.15	1016.15	1016.13	1016.13	1016.12
Is (negative values of $\beta_{opt,s}$ are considered zero)		1016.16	1016.15	1016.13	1016.08	1016.05	1016.07	1015.99	1015.99	1015.91
Total		3900.08	3900.74	3901.14	3902.22	3903.04	3902.66	3904.09	3904.09	3905.09
Total (negative values of $\beta_{opt,s}$ are considered zero)		3900.07	3900.72	3901.11	3902.15	3902.95	3902.58	3903.94	3903.94	3904.88

4.1.4. Semi-annually Optimum Tilt Angle

As mention above, some systems require a high efficiency of the solar system during winter. On the other hand, some systems require a high efficiency of the solar system during summer. Therefore, it is worth finding the semi-annual optimum tilt angle during the warm months and the cold months. The tilt angle is changed once in a period of six months. The warm period is from April to September while the cold period is from October to March. The semi-annual variation of optimum angle $\beta_{opt,sa}$ and the semi-annual total extraterrestrial radiation Isa for different latitude angles and for the main Palestinian cities are listed in Tables 10 and 11.

The results presented in Tables 6 and 7 show that the total yearly radiation received by a surface tilted by $\beta_{opt,s}$ on a seasonal basis is almost the same as that received by a surface $\beta_{opt,sa}$ on a semi-annual basis.

4.1.5. Annually Optimum Tilt Angle

The yearly optimum tilt angles were calculated, and the yearly extraterrestrial radiation I_y gained by a surface tilted by the yearly optimum angle was calculated for different latitudes. The results are listed in Table 12. Similarly, for the main Palestinian cities, the results are listed in Table 13. Table 6 shows the tilt factor on a yearly basis. It could be noted that the increased solar energy on a surface tilted by a yearly optimum angle compared to the horizontal surface increases with increasing latitude of the location. The overall yearly increase of solar radiation is about 1.0% for the latitude of 10.0°, increasing up to 53% for the latitude of 60.0°.

From Table 7, it can be concluded that the increase in solar radiation for a surface tilted by a daily optimum angle relative to a surface tilted by a fixed yearly optimum angle is about 10.6% for most of the Palestinian cities, while the increase in solar radiation for a surface tilted by a monthly optimum angle is about 10.4%. This means that changing the angle each month gives approximately the same increase in solar radiation by changing it daily. Moreover, the increase in solar radiation for a surface tilted by a seasonally optimum angle relative to a surface tilted by a fixed yearly optimum angle is about 8.6% for most of the Palestinian cities, while the increase in solar radiation for a surface tilted by a semi-annually optimum tilt angle is about 8.4%. This means that changing the angle semi-annually gives approximately the same increase in solar radiation by changing it seasonally.

Finally, it can be shown that it is so important to tilt the collector or the panel with the optimum tilt angle. It is clear that for most Palestinian cities, the daily and monthly tilt angle energy gains reached 27% with respect to the horizontal surface, whereas the seasonal and semi-annual tilt angle energy gains reached 25%. Lastly, the energy gains for the yearly tilt angle were about 15% for most of the Palestinian cities.

4.2. Part Two: Results Based on PVWatts and PVGIS

Part one results showed a small variation between Palestinian cities by using either the optimum tilt angles as one approach or the amount of total radiation as the second approach. For that reason, two major cities in Palestine are selected; the first city is Jerusalem which is located in the West Bank and the other is Gaza city located in the Gaza Strip.

Jerusalem is located in the West Bank where there is a meteorological station; PVWatts depends on this station for global radiation. In Gaza city which is located in the Gaza Strip, the nearest meteorological station is Be'Er Sheva.

The reason for using PVWatts and PVGIS for this comparison is that PVWatts uses solar resource data collected from a ground station whereas PVGIS uses data produced from satellite images.

PVWatts does not estimate the optimum tilt angle, whereas PVGIS estimates only the yearly optimum tilt angle. Therefore, the optimum tilt angle for each city and particular time period is found by estimating PV outputs with various tilt angles for this particular period so the optimum tilt angle giving the maximum output is found.

Table 10. The semi-annual optimum tilt angle $\beta_{opt,sa}$ and extraterrestrial radiation I_{sa} in kWh/m^2 for different latitudes.

Period		Latitude												
		0	5	10	15	20	25	30	35	40	45	50	55	60
Warm period (April to September)	$\beta_{opt,sa}$	−22.26	−17.65	−13.05	−8.48	−3.95	0.52	4.93	9.23	13.40	17.34	20.95	23.97	25.82
	I_{sa}	1937.51	1942.25	1946.01	1948.80	1950.53	1951.05	1950.06	1947.10	1941.36	1931.50	1915.14	1887.75	1839.69
Isa (negative values of $\beta_{opt,sa}$ are considered zero)		1793.09	1850.84	1895.74	1927.47	1945.89	1951.05	1950.06	1947.10	1941.36	1931.50	1915.14	1887.75	1839.69
Cold period (October to March)	$\beta_{opt,sa}$	22.59	27.21	31.84	36.46	41.08	45.68	50.26	54.81	59.31	63.75	68.09	72.28	76.20
	I_{sa}	2015.31	2009.12	2001.70	1992.82	1982.15	1969.21	1953.27	1933.24	1907.43	1873.02	1825.09	1754.11	1638.58
Total		3952.82	3951.37	3947.71	3941.62	3932.68	3920.26	3903.33	3880.34	3848.79	3804.52	3740.23	3641.86	3478.27
Total (negative values of $\beta_{opt,sa}$ are considered zero)		3808.40	3859.96	3897.44	3920.29	3928.04	3920.26	3903.33	3880.34	3848.79	3804.52	3740.23	3641.86	3478.27

Table 11. The semi-annual optimum tilts angle $\beta_{opt,sa}$ and extraterrestrial radiation I_{sa} in kWh/m^2 for the main Palestinian cities.

Period		City								
		Jenin	Tulkarm	Nablus	Ramallah	Jerusalem	Jericho	Hebron	Gaza	Rafah
Warm period (April to September)	$\beta_{opt,sa}$	7.06	6.93	6.86	6.58	6.48	6.55	6.26	6.26	6.05
	I_{sa}	1948.89	1948.98	1949.03	1949.20	1949.27	1949.22	1949.40	1949.40	1949.51
Cold period (October to March)	$\beta_{opt,sa}$	52.51	52.37	52.29	52.00	51.89	51.97	51.66	51.66	51.44
	I_{sa}	1944.01	1944.61	1944.96	1946.22	1946.68	1946.33	1947.64	1947.64	1948.55
Total		3892.90	3893.59	3893.99	3895.42	3895.95	3895.55	3897.04	3897.04	3898.06

Table 12. The annual optimum tilt angle $\beta_{opt,y,}$ and extraterrestrial radiation I_y in kWh/m^2 for a different latitude.

	Latitude												
	0	5	10	15	20	25	30	35	40	45	50	55	60
$\beta_{opt,y}$	0.63	5.18	9.73	14.25	18.75	23.21	27.62	31.94	36.14	40.17	43.92	47.18	49.45
I_y	3653.99	3652.54	3648.73	3642.33	3632.91	3619.77	3601.84	3577.44	3543.91	3496.82	3428.34	3323.42	3148.68

Table 13. The annual optimum tilt angle $\beta_{opt,y}$ and extraterrestrial radiation I_y in kWh/m^2 for the main Palestinian cities.

	City								
	Jenin	**Tulkarm**	**Nablus**	**Ramallah**	**Jerusalem**	**Jericho**	**Hebron**	**Gaza**	**Rafah**
$\beta_{opt,y}$	29.75	29.62	29.55	29.27	29.17	29.24	28.95	28.95	28.74
I_y	3590.78	3591.50	3591.93	3593.44	3594.00	3593.58	3595.16	3595.16	3596.25

The optimum tilt angles for two Palestinian cities, Jerusalem and Gaza city, are shown in Table 14 for the different time periods. Figures 9 and 10 show a comparison of monthly and seasonal optimum tilt angles for Jerusalem and Gaza using PVWatts and PVGIS with those calculated by the mathematical model. It is clear that the results are very close. This means that the mathematical model in part one gives accurate results for the optimum tilt angle.

Table 14. A comparison of monthly, seasonal, semi-annual, and yearly optimum tilt angles (degrees) for Jerusalem and Gaza City based on PVWatts and PVGIS with those calculated in the present work.

	Jerusalem			**Gaza City**		
Period	**Yearly Optimum Tilt Angle**					
	Mathematical Model	**PVGIS**	**PVWatts**	**Mathematical Model**	**PVGIS**	**PVWatts**
	29	28	26	29	28	29
	Semi-annual optimum tilt angle					
Warm period (April to September)	6	9	11	6	9	12
Cold period (October to March)	52	51	46	52	50	50
	Seasonal optimum tilt angle					
Autumn	44	43	41	44	43	43
Winter	57	56	51	57	55	56
Spring	17	20	17	16	19	19
Summer	−1	3	7	−1	0	5
	Monthly optimum tilt angle					
January	59	58	55	59	59	59
February	50	50	49	50	50	50
March	35	35	32	35	35	35
April	16	16	17	16	17	16
May	0	1	4	0	2	6
June	−7	0	0	−8	0	0
July	−4	0	2	−4	0	0
August	10	11	14	9	10	12
September	29	29	29	28	29	29
October	46	46	46	46	46	46
November	57	57	56	57	57	57
December	62	61	56	62	61	61

It worth mentioning that PVGIS and PVWatts do not estimate a negative tilt angle. This means the minimum tilt angle is zero (horizontal surface).

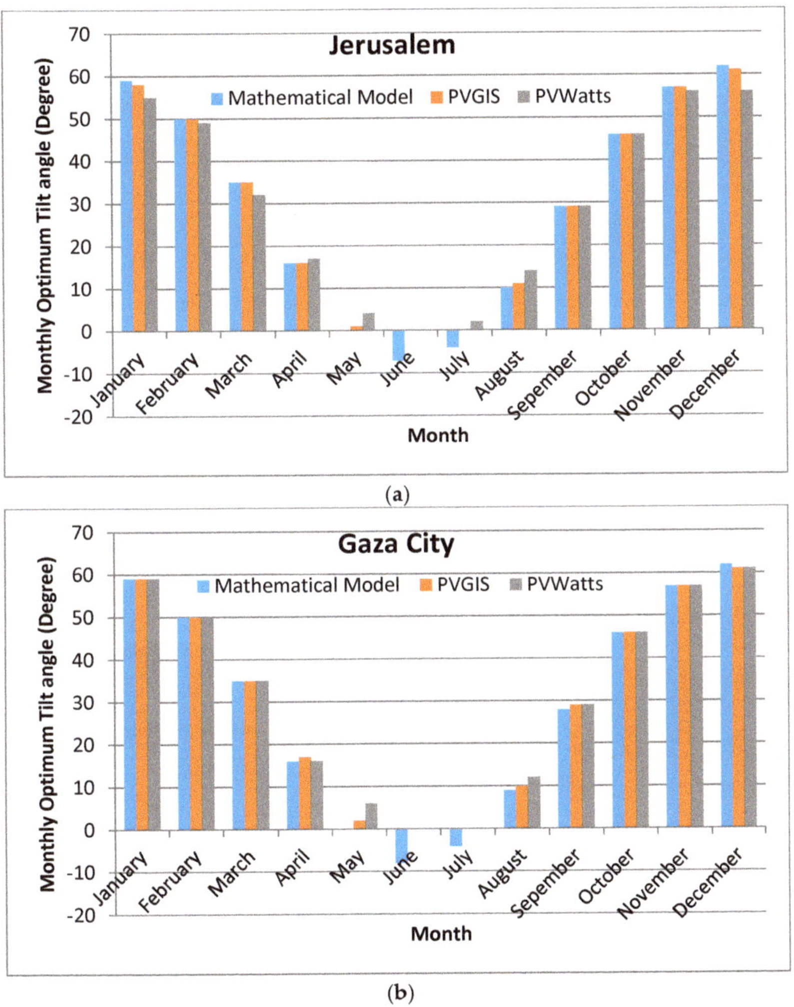

Figure 9. A comparison of monthly optimum tilt angles based on PVWatts and PVGIS with those calculated by the mathematical model: Jerusalem (**a**), Gaza City (**b**).

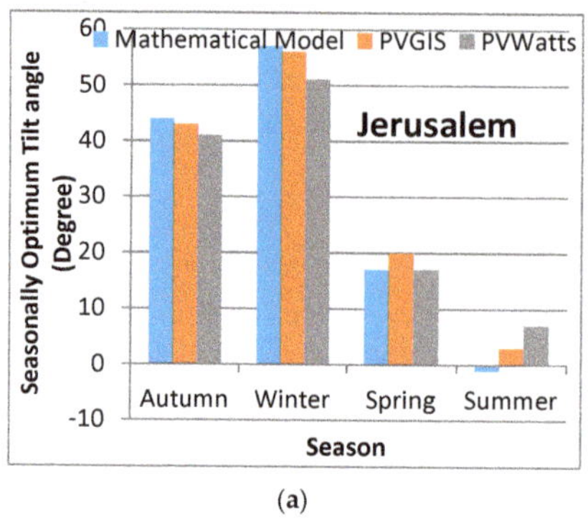

(a)

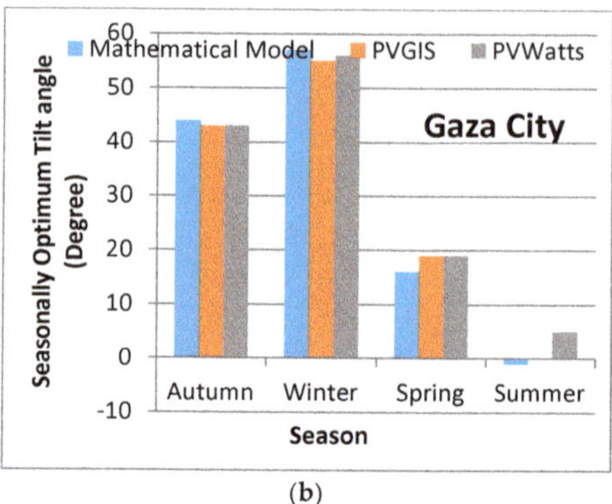

(b)

Figure 10. A comparison of seasonal optimum tilt angles based on PVWatts and PVGIS with those calculated by the mathematical model: Jerusalem (**a**), Gaza city (**b**).

5. Case Studies of two Palestinian Cities

To perform this study, the cities of Jerusalem and Gaza city have been selected. Jerusalem is located in the West Bank whereas Gaza city is located in the Gaza Strip.

PVGIS and PVWatts are used to calculate the energy generated in the two locations mentioned above for a photovoltaic composed of a monocrystalline silicon module with 5 kWP of peak power installed. The comparison is made with five configurations: a system with a PV in a horizontal position and annual, semi-annual, seasonal, and monthly optimum tilt angle positions. In this way, the differences and gains due to the optimal position can be quantitatively verified.

Firstly, PVGIS and PVWatts are used to obtain the monthly average global radiation for Jerusalem and Gaza. Tables 15 and 16 show the monthly average global radiation in Jerusalem and Gaza city for different tilt angles (β).

Table 15. The monthly average global radiation in Jerusalem for different tilt angles (β).

Months	Monthly Average Radiation (kWh/m²/day) for Jerusalem											
	PVGIS						PVWatts					
	β (0°)	β (28°)	β (29°)	β (30°)	β (60°)	β (90°)	β (0°)	β (26°)	β (29°)	β (30°)	β (60°)	β (90°)
January	3.07	4.35	4.39	4.42	4.84	4.16	2.68	3.57	3.64	3.66	3.88	3.27
February	3.86	4.96	4.96	5.00	5.11	4.11	3.57	4.49	4.56	4.57	4.61	3.66
March	5.19	6.00	6.00	6.00	5.58	3.97	4.81	5.43	5.45	5.45	4.95	3.44
April	6.40	6.60	6.57	6.57	5.43	3.20	6.13	6.37	6.33	6.32	5.16	2.98
May	7.19	6.81	6.77	6.74	4.97	2.36	7.17	6.93	6.83	6.79	5.01	2.43
June	8.17	7.40	7.33	7.27	5.00	1.91	8.07	7.53	7.38	7.33	5.03	2.08
July	8.06	7.42	7.39	7.32	5.16	2.12	8.01	7.61	7.47	7.43	5.22	2.23
August	7.39	7.39	7.35	7.35	5.74	2.98	7.28	7.45	7.39	7.36	5.79	3.04
September	6.20	6.97	6.97	6.97	6.20	4.07	6.26	7.08	7.1	7.1	6.32	4.13
October	4.65	5.84	5.87	5.90	5.87	4.52	4.69	5.86	5.94	5.96	5.88	4.49
November	3.53	4.97	5.00	5.03	5.47	4.63	3.34	4.54	4.63	4.66	4.94	4.12
December	2.86	4.26	4.29	4.32	4.87	4.29	2.45	3.38	3.46	3.48	3.77	3.24
Annual Average	5.55	6.08	6.07	6.07	5.35	3.53	5.37	5.85	5.85	5.84	5.05	3.26

The maximum radiation value is reached in the horizontal PV in June for the two cities and using PVGIS and PVWatts, whereas the maximum monthly radiation for a PV tilted at an annual optimum tilt angle is achieved in July. It can be concluded that the annual average radiation (kWh/m²/day) for Gaza city is slightly higher than that for Jerusalem. Moreover, the annual average radiation (kWh/m²/day) calculated by using PVGIS is (6.12) slightly higher than the annual average radiation (5.96) calculated using PVWatts.

Table 16. The monthly average global radiation in Gaza city for different tilt angles (β).

Months	Monthly Average Radiation (kWh/m^2/day) for Gaza city										
	PVGIS						PVWatts				
	β (0°)	β (28°)	β (29°)	β (30°)	β (60°)	β (90°)	β (0°)	β (29°)	β (30°)	β (60°)	β (90°)
January	3.02	4.29	4.32	4.35	4.77	4.10	3.05	4.53	4.57	5.04	4.34
February	3.96	5.11	5.14	5.18	5.29	4.25	3.72	4.93	4.96	5.07	4.04
March	5.32	6.19	6.19	6.19	5.77	4.06	4.97	5.8	5.81	5.37	3.76
April	6.53	6.77	6.73	6.73	5.53	3.22	6.19	6.4	6.39	5.21	3.01
May	7.29	6.90	6.87	6.84	5.03	2.33	7.08	6.74	6.7	4.93	2.39
June	8.07	7.30	7.23	7.20	4.90	1.86	7.84	7.11	7.06	4.85	2.04
July	7.94	7.32	7.26	7.23	5.06	2.07	7.55	7.00	6.95	4.93	2.2
August	7.32	7.32	7.29	7.26	5.68	2.92	7.03	7.06	7.03	5.51	2.92
September	6.17	6.90	6.93	6.93	6.17	4.03	5.84	6.58	6.58	5.84	3.82
October	4.81	6.10	6.13	6.13	6.13	4.71	4.56	5.89	5.91	5.91	4.54
November	3.50	4.97	5.00	5.03	5.47	4.63	3.63	5.08	5.11	5.44	4.54
December	2.89	4.32	4.39	4.42	4.97	4.35	2.84	4.43	4.47	5.05	4.44
Annual Average	5.57	6.12	6.12	6.12	5.40	3.55	5.36	5.96	5.96	5.26	3.50

It can be noticed that annual average radiation (kWh/m^2/day) for a PV tilted at the yearly optimum tilt angle calculated by the mathematical model suggested in this study and the annual average radiation (kWh/m^2/day) for a PV tilted at the yearly optimum tilt angle calculated by PVGIS and PVWatts is very similar. For example, the yearly optimum tilt angle calculated by the suggested mathematical model in this study for Jerusalem is 29° and that when using PVWatts is 26° as shown in Table 15. However, the annual average radiation is the same (5.85 kWh/m^2/day) as shown in Table 16.

In another example, the yearly optimum tilt angle calculated by the suggested mathematical model in this study for Gaza is 29°and that when using PVGIS is 28°as shown in Table 16. However, the annual average radiation is the same (6.12 kWh/m^2/day) as shown in Table 16.

For the city of Jerusalem and using PVGIS, it is observed that the annual average radiation on a horizontal plane is (5.55 kWh/m^2/day) as shown in Table 15 while for 28°(PVGIS yearly optimum tilt angle), it is (6.08 kWh/m^2/day) as shown in Table 15, and for 29°(yearly optimum tilt angle obtained from the mathematical model), it is (6.07 kWh/m^2/day) as shown in Table 15. It can be concluded that increasing the tilt angle more than the optimal value also leads to a decrease in the annual average radiation, obtaining only (3.53 kWh/m^2/day) as shown in Table 16 for a vertical plane β (90°).

For the city of Gaza and using PVWatts, it is observed that the annual average radiation on a horizontal plane is (5.36 kWh/m^2/day) as shown in Table 16 while for 29°(the same yearly optimum tilt angle using PVWatts and the mathematical model), it is (5.96 kWh/m^2/day) as shown in Table 16. For a vertical plane, it is only (3.5 kWh/m^2/day) as shown in Table 16.

For a photovoltaic system with a peak power of (5 kWh) in Jerusalem, the following monthly values are calculated using PVGIS and PVWatts.

Tables 17 and 18 show the monthly energy generated in Jerusalem and Gaza city by a 5 kWh system.

Figures 11 and 12 show the monthly energy generated in Jerusalem and Gaza city by a 5 kWh system.

It can be concluded that the energy generated by a PV system fixed at a yearly optimum tilt angle is more than 10% higher than that for a horizontal PV system. Moreover, the energy generated by a PV system fixed at semi-annual tilt angle gives 15% more energy gain as shown in Table 19.

Table 17. Monthly energy generated by a 5 kWh system in Jerusalem.

Months	Monthly Energy Generated (kWh) For Jerusalem									
	PVGIS					PVWatts				
	$\beta = 0$	$\beta_{opt,y}$	$\beta_{opt,\,sa}$	$\beta_{opt,\,s}$	$\beta_{opt,\,m}$	$\beta = 0$	$\beta_{opt,y}$	$\beta_{opt,\,sa}$	$\beta_{opt,\,s}$	$\beta_{opt,\,m}$
January	382	553	612	614	617	339	457	496	498	498
February	431	556	584	580	584	410	517	540	539	540
March	632	726	708	712	730	602	676	663	663	679
April	732	751	752	758	758	721	746	746	751	751
May	832	782	830	810	833	844	813	847	838	848
June	896	804	881	893	896	918	853	909	915	918
July	908	832	899	907	908	933	883	931	935	935
August	841	837	856	848	857	845	861	869	864	872
September	697	776	738	761	776	706	793	759	784	795
October	548	689	709	713	713	557	692	718	718	718
November	413	592	651	640	653	393	534	580	573	582
December	352	539	611	616	617	306	426	469	473	474
Annual Sum	7664	8437	8831	8852	8942	7574	8251	8527	8551	8610
Percentage gain with respect to a horizontal plane (%)		10.1	15.2	15.5	16.7		8.9	12.6	12.9	13.7

Table 18. Monthly energy generated by a 5 kWh system in Gaza City.

Months	Monthly Energy Generated (kWh) for Gaza City									
	PVGIS					PVWatts				
	$\beta = 0$	β opt,y	$\beta_{opt,\,sa}$	$\beta_{opt,\,s}$	$\beta_{opt,\,m}$	$\beta = 0$	$\beta_{opt,y}$	$\beta_{opt,\,sa}$	$\beta_{opt,\,s}$	$\beta_{opt,\,m}$
January	369	535	592	594	594	371	557	614	618	618
February	438	571	601	598	601	408	540	565	561	565
March	656	758	743	740	763	610	709	692	691	712
April	771	794	793	801	801	703	723	729	732	732
May	879	827	878	859	880	820	776	821	809	824
June	926	833	912	926	926	867	784	852	864	867
July	935	857	926	935	935	861	795	853	861	861
August	857	851	871	857	872	796	796	816	808	816
September	704	785	745	769	785	648	726	698	710	726
October	568	722	747	749	749	527	679	699	701	701
November	406	583	641	631	644	418	586	629	622	629
December	347	536	608	614	615	342	543	610	617	618
Annual Sum	7856	8652	9057	9073	9165	7371	8214	8578	8594	8669
Percentage gain with respect to a horizontal plane (%)		10.1	15.3	15.5	16.7		11.4	16.4	16.6	17.6

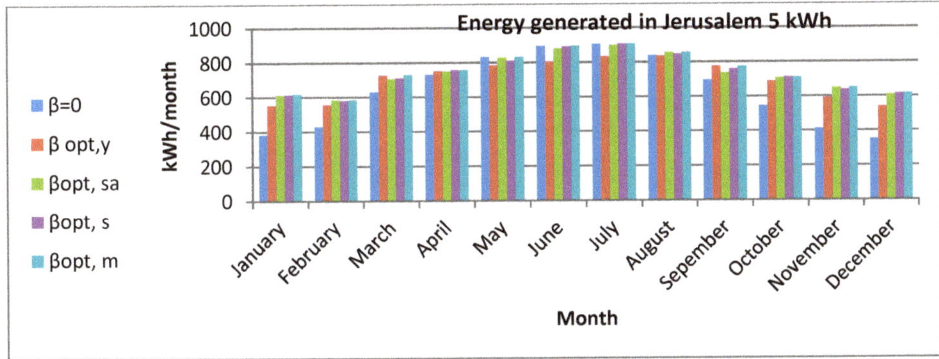

Figure 11. Monthly energy generated in Jerusalem by a 5 kWh system.

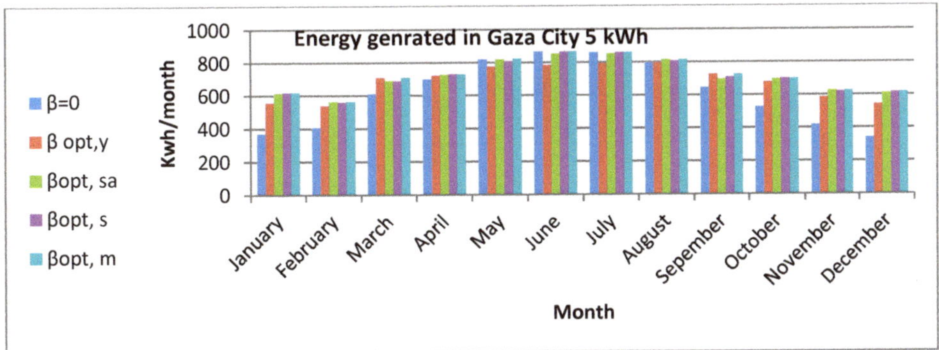

Figure 12. Monthly energy generated in Gaza city by a 5 kWh system.

It can be noted that a slight energy gain can be obtained by a seasonally optimum tilt angle compared to a semi-annual optimum tilt angle.

Changing the tilt angle 12 times through the year to obtain the monthly optimum tilt leads to about 17% energy gain with respect to a horizontal PV system as shown in Tables 17 and 18. This means that it is worth fixing the PV system at the yearly optimum tilt angle and changing the optimum tilt angle at least two times per year.

6. Comparison of the Present Work with Previously Published Papers

Firstly, a comparison of the present mathematical model was made with H. Darhmaoui and D. Lahjouji [37]. They used a mathematical model to estimate the solar radiation on a tilted surface at 35 sites in different countries of the Mediterranean region. They found the yearly optimum tilt angle for Gaza city to be 32°, whereas in the present mathematical model the value was 29°.

Secondly, a comparison of the present mathematical model with the models of Nijegorodor et al. [39] and P. Talebizadeh et al. [19] was made. Nijegorodor et al. and P. Talebizadeh et al. suggested monthly and yearly equations for calculating monthly and yearly optimum tilt angles. Monthly and yearly equations were used to estimate the optimum monthly and yearly tilt angles for Jerusalem, and a comparison of these angles with those obtained by the present mathematical model is given in Table 19 and is plotted in Figure 13. Results show that the yearly optimum tilt angles are very close while the monthly optimum tilt angle is slightly different.

Table 19. A comparison of monthly and yearly tilt angles for Jerusalem based on Nijegorodor et al. and P. Talebizadeh et al. with the values calculated in the present mathematical model.

Month		Present Mathematical Model	Nijegorodor et al. [39]	P. Talebizadeh et al. [31]
January	$\beta_{opt,m}$	59.28	57.28	56.10
February	$\beta_{opt,m}$	50.44	47.83	47.30
March	$\beta_{opt,m}$	35.21	35.78	31.59
April	$\beta_{opt,m}$	15.81	21.78	16.40
May	$\beta_{opt,m}$	−0.21	5.56	2.74
June	$\beta_{opt,m}$	−7.31	−6.35	−2.17
July	$\beta_{opt,m}$	−3.94	−1.72	0.16
August	$\beta_{opt,m}$	9.53	13.83	12.09
September	$\beta_{opt,m}$	28.66	29.78	28.70
October	$\beta_{opt,m}$	46.15	43.78	44.03
November	$\beta_{opt,m}$	57.41	54.56	54.89
December	$\beta_{opt,m}$	61.74	61.65	58.54
year	$\beta_{opt,m}$	29.17	30.31	28.83

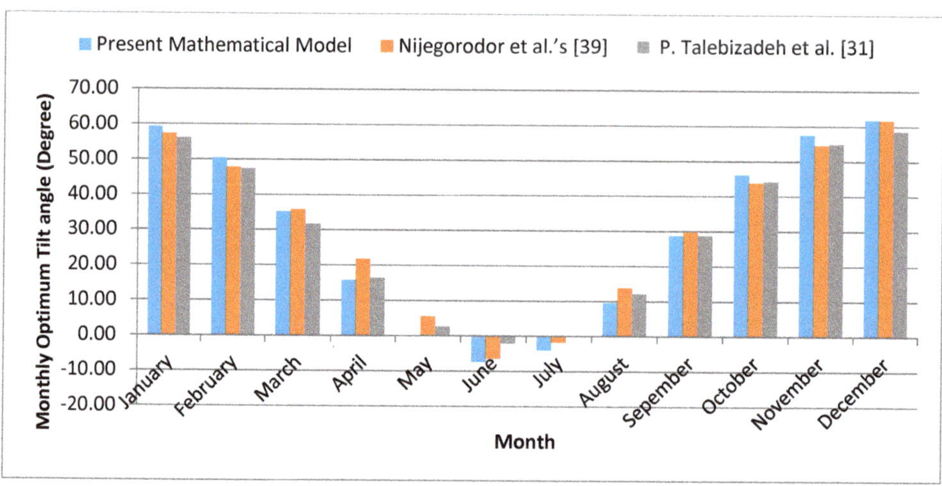

Figure 13. Comparison of monthly and yearly tilt angles for Jerusalem based on Nijegorodor et al [39]. and P. Talebizadeh et al [31], with values calculated in the present mathematical model.

7. Conclusions

It is clear that the suggested mathematical model gives accurate results for the optimum tilt angle compared to PVWatts and PVGIS. It is worth designing a solar panel or collector with a negative tilt angle for locations at latitudes from 0° to 30°. For all the Palestinian cities, there is no need to design a collector with a negative tilt angle. Compared to a horizontal surface, yearly, semi-annual, seasonal, and monthly adjustments of solar panels in the main Palestinian cities will generate approximately 10%, 15%, 15%, and 17% more energy, respectively, whereas compared to the fixed optimum yearly tilt angle, monthly adjustments of the solar panels in the main Palestinian cities can generate about 6% more energy. Seasonal and semi-annual adjustments can generate about 5% more solar energy. This shows that solar panels or collectors have to be adjusted monthly or semi-annually. For the roof top-solar PV rated power (5 kWP) which was taken as a case study for the present work, it is convenient to change it semi-annually. Moreover, this proves the importance of an accurate tilt angle for the solar panels or collectors. If a supporting structure for the PV is well designed, then the solar PV can be easily adjusted in a limited time and can be performed by an unqualified person.

The daily optimum tilt angle for the main Palestinian cities ranges from −8.0° in June to 62° in December. The monthly optimum tilt angle for the main Palestinian cities throughout the year is between −7.0° in June and 62° in December. The seasonal optimum tilt angle for the main Palestinian cities throughout the year is between a minimum value of −1.0° in June and a maximum value of 58° in December. Moreover, the optimum tilt angle for autumn is more than the latitude of the city (44°) and for spring, less than the latitude of the city (17°). The semi-annual optimum tilt angle for the main Palestinian cities throughout the year is between 7.0° in the warm period (April to September) and 52° in the cold period (October to March).

The yearly average radiation in Jerusalem using PVGIS is (5.55 kWh/m^2/day) based on a horizontal surface while using a panel tilted with a yearly optimum tilt angle, the value is (6.08 kWh/m^2/day). The yearly average radiation in Jerusalem using PVWatts is (5.37 kWh/m^2/day) based on a horizontal surface while using a panel tilted with a yearly optimum tilt angle, the value is (5.85 kWh/m^2/day). The yearly average radiation in Gaza using PVGIS is (5.57 kWh/m^2/day) based on a horizontal surface while using a panel tilting with a yearly optimum tilt angle, the value is (6.12 kWh/m^2/day). The yearly average radiation in Jerusalem using PVWatts is (5.36 kWh/m^2/day) based on a horizontal surface while using a panel tilted with a yearly optimum tilt angle, the value is (5.96 kWh/m^2/day). It is clear that the PVGIS results are slightly higher than the PVWatts results for Palestinian cities.

The yearly optimum tilt angle for most Palestinian cities is about 29°. It can be concluded that the monthly optimum tilt angle for a particular month is approximately the average of the daily optimum tilt angle for the particular month, while the seasonal, semi-annual, and yearly optimum tilt angles are approximately the average monthly optimum tilt angle for that particular period.

Author Contributions: R.A., and A.J. conceived of the idea and wrote the article; A.J. and R.A. analyzed the data; R.A., A.J., and S.A.-F. wrote the paper. F.M.-A. supervised the research and revised the manuscript. All authors contributed to the structure and aims of the manuscript, paper drafting, editing, and review. All authors have read and approved the final manuscript.

Funding: This reseaech received no external funding.

Acknowledgments: The authors would like to acknowledge the Palestinian Central Bureau of Statistics (Eng. Haitham Zeidan) and An Najah National University for facilitating this research.

Conflicts of Interest: The authors declare no conflict of interest.

Abbreviations

δ	Declination angle
n	The day number
$\emptyset$	Latitude angle
h	Hour angle
α	The altitude angle
h_s	The hour angles at sunrise and sunset
β	The tilt angle
ASHRAE	American Society of Heating, Refrigerating and Air-Conditioning Engineers
I_d	Daily extraterrestrial radiation
I_p	Extraterrestrial radiation for a particular period
PV	Photovoltaic
$\beta_{opt,d}$	Daily optimum tilt angle
$\beta_{opt,m}$	Monthly optimum tilt angle
I_m	Monthly total extraterrestrial radiation
F	The tilt factor
$\beta_{opt,s}$	Seasonally optimum tilt angle

I_s	Seasonal extraterrestrial radiation
$\beta_{opt,sa}$	Semi-annual optimum angle
I_{sa}	Semi-annal extraterrestrial radiation
I_y	The yearly extra-terrestrial radiation
$\beta_{opt,y}$	The annual optimum tilt angle
PEA	Palestinian Energy Authority
PCBS	Palestinian Central Bureau of Statistics
NREL	National Renewable Energy Laboratory -USA

References

1. Juaidi, A.; Montoya, F.G.; Ibrik, I.H.; Manzano-Agugliaro, F. An overview of renewable energy potential in Palestine. *Renew. Sust. Energ. Rev.* **2016**, *65*, 943–960. [CrossRef]
2. Mohammed, M. (Palestine Investment Fund, Palestine). Palestine's Oil and Gas Resources: Prospects and Challenges. 2014. Available online: http://thisweekinpalestine.com/wp-content/uploads/2014/07/Palestine%E2%80%99s-Oil-and-Gas.pdf (accessed on 28 January 2020).
3. European Joint Strategy in Support of Palestine. *Towards a Democratic and Accountable Palestinian state (2017–2020)—European Joint Programming*; European Union: Brussels, Belgium; p. 64. Available online: https://eeas.europa.eu/sites/eeas/files/final_-_european_joint_strategy_english.pdf (accessed on 28 January 2020).
4. West Bank and Gaza Energy Efficiency Action Plan 2020–2030. Final Report, ACS19044; World Bank Group: Energy and Extractives. Energy Sector Management Program (ESMAP): Gaza, Palestine, 2016; p. 67. Available online: http://documents.worldbank.org/curated/en/851371475046203328/pdf/ACS19044-REPLACEMENT-PUBLIC-FINAL-REPORT-P147961-WBGaza-Energy-Efficiency-Action-Plan.pdf (accessed on 30 January 2020).
5. Ibrik, I.H.; Mahmoud, M.M. Energy efficiency improvement procedures and audit results of electrical, thermal and solar applications in Palestine. *Energy Policy* **2005**, *33*, 651–658. [CrossRef]
6. Hamed, T.A.; Flamm, H.; Azraq, M. Renewable energy in the Palestinian Territories: Opportunities and challenges. *Renew. Sust. Energ. Rev.* **2012**, *16*, 1082–1088. [CrossRef]
7. Ismail, M.S.; Moghavvemi, M.; Mahlia, T.M.I. Design of an optimized photovoltaic and microturbine hybrid power system for a remote small community: Case study of Palestine. *Energy Convers. Manag.* **2013**, *75*, 271–281. [CrossRef]
8. *Status of The Environment in The State of Palestine 2015*; Applied Research Institute Jerusalem (ARIJ); Swiss Agency for Development and Cooperation SDC, 2015; p. 189. Available online: http://www.arij.org/ (accessed on 28 January 2020).
9. Palestinian Central Bureau of Statistics (PCBS). Available online: http://www.pcbs.gov.ps/ (accessed on 28 January 2020).
10. *Palestinian National Plan 2011–13*; Energy Sector Strategy; Ministry of Planning an Administrative Development Palestine: Ramallah, Palestine, 2011. Available online: https://www.google.com/url?sa=t&rct=j&q=&esrc=s&source=web&cd=4&ved=2ahUKEwim8rWjtqjnAhXR4KQKHel6A10QFjADegQIAhAB&url=http%3A%2F%2Fwww.lacs.ps%2FdocumentsShow.aspx%3FATT_ID%3D4799&usg=AOvVaw0ih44V_42LHzdNHvQp57_l (accessed on 30 January 2020).
11. El-Kassaby, M.M. Monthly and daily optimum tilt angle for south facing solar collectors; theoretical model, experimental and empirical correlations. *Sol. Wind Technol.* **1998**, *5*, 589–596. [CrossRef]
12. Duffie, J.A.; Beckman, W.A. *Solar Engineering of Thermal Processes*, 2nd ed.; John Wiley and Sons: New York, NY, USA, 1991; p. 919.
13. Lunde, P.J. *Solar Thermal Engineering*; John Wiley and Sons: New York, NY, USA, 1980; p. 635.
14. Jafarkazemi, F.; Saadabadi, S.A. Optimum tilt angle and orientation of solar surfaces in Abu Dhabi, UAE. *Renew. Energy* **2013**, *56*, 44–49. [CrossRef]
15. Pavlović, T.; Pavlović, Z.; Pantić, L.; Kostić, L. Determining optimum tilt angles and orientations of photovoltaic panels in niš, Serbia. *Contemp. Mater. I* **2010**, 151–156. [CrossRef]
16. Omar, M.A.; Mahmoud, M.M. Grid connected PV-home systems in Palestine: A review on technical performance, effects and economic feasibility. *Renew. Sust. Energy Rev.* **2018**, *82*, 2490–2497. [CrossRef]

17. National Renewable Energy Laboratory (NREL). PV Watts Calculator. Available online: https://pvwatts.nrel.gov/ (accessed on 30 January 2020).
18. Photovoltaic Geographical Information System. European Commission. PVGIS. Available online: https://re.jrc.ec.europa.eu/pvg_tools/en/tools.html (accessed on 28 January 2020).
19. Berisha, X.; Zeqiri, A.; Meha, D. Solar Radiation—The Estimation of the Optimum Tilt Angles for South-Facing Surfaces in Pristina. *Preprints* **2017**, 13. [CrossRef]
20. Racharla, S.; Rajan, K. Solar tracking system—A review. *Int. J. Sustain. Eng.* **2017**, *10*, 72–81.
21. Mousazadeh, H.; Keyhani, A.; Javadi, A.; Mobli, H.; Abrinia, K.; Sharifi, A. A review of principle and sun-tracking methods for maximizing solar systems output. *Renew. Sust. Energ. Rev.* **2009**, *13*, 1800–1818. [CrossRef]
22. Ahmad, M.J.; Tiwari, G.N. Optimization of tilt angle for solar collectors to receive maximum radiation. *Open Renew. Energy J.* **2009**, *2*, 19–24.
23. Abdulsalam, H.S.; Alibaba, H.Z. Optimum Tilt Angle for Photovoltaic Panels in Famagusta, Cyprus. *IJSSHR* **2019**, *7*, 29–35.
24. Jiménez-Torres, M.; Rus-Casas, C.; Lemus-Zúiga, L.; Hontoria, L. The Importance of Accurate Solar Data for Designing Solar Photovoltaic Systems—Case Studies in Spain. *Sustainability* **2017**, *9*, 247.
25. Mousavi Maleki, S.A.; Hizam, H.; Gomes, C. Estimation of hourly, daily and monthly global solar radiation on inclined surfaces: Models re-visited. *Energies* **2017**, *10*, 134. [CrossRef]
26. Skeiker, K. Optimum tilt angle and orientation for solar collectors in Syria. *Energy Convers. Manag.* **2009**, *50*, 2439–2448. [CrossRef]
27. Khasawneh, Q.A.; Damra, Q.A.; Salman, B.; Husni, O. Determining the Optimum Tilt Angle for Solar Applications in Northern Jordan. *JJMIE* **2015**, *9*, 187–193.
28. Hartner, M.; Ortner, A.; Hiesl, A.; Haas, R. East to west—The optimal tilt angle and orientation of photovoltaic panels from an electricity system perspective. *Appl. Energy* **2015**, *160*, 94–107. [CrossRef]
29. Safdarian, F.; Nazari, M.E. Optimal tilt angle and orientation for solar collectors in Iran. In Proceedings of the IEEE 10th International Symposium on Diagnostics for Electrical Machines, Power Electronics and Drives (SDEMPED), Guarda, Portugal, 1–4 September 2015; IEEE: Guarda, Portugal, 2015; pp. 494–500.
30. Jacobson, M.Z.; Jadhav, V. World estimates of PV optimal tilt angles and ratios of sunlight incident upon tilted and tracked PV panels relative to horizontal panels. *Sol. Energy* **2018**, *169*, 55–66. [CrossRef]
31. Talebizadeh, P.; Mehrabian, M.A.; Abdolzadeh, M. Determination of optimum slope angles of solar collectors based on new correlations. *Energy Sources Part A Recovery Util. Environ. Eff.* **2011**, 33, 1567–1580. [CrossRef]
32. Morcos, V.H. Optimum tilt angle and orientation for solar collectors in Assiut, Egypt. *Renew. Energy* **1994**, *4*, 291–298. [CrossRef]
33. Hartley, L.E.; Martínez-Lozano, J.A.; Utrillas, M.P.; Tena, F.; Pedros, R. The optimisation of the angle of inclination of a solar collector to maximise the incident solar radiation. *Renew. Energy* **1999**, *17*, 291–309. [CrossRef]
34. Yakup, M.A.B.H.M.; Malik, A.Q. Optimum tilt angle and orientation for solar collector in Brunei Darussalam. *Renew. Energy* **2001**, *24*, 223–234. [CrossRef]
35. Tang, R.; Wu, T. Optimal tilt-angles for solar collectors used in China. *Appl. Energy* **2004**, *79*, 239–248. [CrossRef]
36. Ulgen, K. Optimum tilt angle for solar collectors. *Energy Sources Part A Recovery Util. Environ. Eff.* **2006**, *28*, 1171–1180. [CrossRef]
37. Darhmaoui, H.; Lahjouji, D. Latitude based model for tilt angle optimization for solar collectors in the Mediterranean region. *Energy Procedia* **2013**, *42*, 426–435. [CrossRef]
38. Guo, M.; Zang, H.; Gao, S.; Chen, T.; Xiao, J.; Cheng, L.; Sun, G. Optimal tilt angle and orientation of photovoltaic modules using HS algorithm in different climates of China. *Appl. Sci.* **2017**, *7*, 1028. [CrossRef]
39. Nijegorodov, N.; Devan, K.R.S.; Jain, P.K.; Carlsson, S. Atmospheric transmittance models and an analytical method to predict the optimum slope of an absorber plate, variously oriented at any latitude. *Renew. Energy* **1994**, *4*, 529–543. [CrossRef]

© 2020 by the authors. Licensee MDPI, Basel, Switzerland. This article is an open access article distributed under the terms and conditions of the Creative Commons Attribution (CC BY) license (http://creativecommons.org/licenses/by/4.0/).

Article

A Rebalancing Strategy for the Imbalance Problem in Bike-Sharing Systems

Peiyu Yi, Feihu Huang and Jian Peng *

College of Computer Science, Sichuan University, Chengdu 610065, Sichuan, China
* Correspondence: jianpeng@scu.edu.cn; Tel.: +86-28-8546-9720

Received: 25 May 2019; Accepted: 1 July 2019; Published: 4 July 2019

Abstract: Shared bikes have become popular traveling tools in our daily life. The successful operation of bike sharing systems (BSS) can greatly promote energy saving in a city. In BSS, stations becoming empty or full is the main cause of customers failing to rent or return bikes. Some truck-based rebalancing strategies are proposed to solve this problem. However, there are still challenges around the relocation of bikes. The truck operating costs also need to be considered. In this paper, we propose a customer-oriented rebalancing strategy to solve this problem. In our strategy, two algorithms are proposed to ensure the whole system is balanced for as long as possible. The first algorithm calculates the optimal state of each station through the one-dimensional Random Walk Process with two absorption walls. Based on the derived optimal state of each station, the second algorithm recommends the station that has the largest difference between its current state and its optimal state to the customer. In addition, a simulation system of shared bikes based on the historical records of Bay Area Bikeshare is built to evaluate the performance of our proposed rebalancing strategy. The simulation results indicate that the proposed strategy is able to effectively decrease the imbalance in the system and increase the system's performance compared with the truck-based methods.

Keywords: bike sharing; energy saving; system rebalancing

---

1. Introduction

Energy saving is a vital task for the sustainable development of a city [1]. The government attaches great importance to the development of bike sharing systems (BSS), since they can greatly help reduce urban energy consumption [2,3]. We think the essential condition for the successful operation of BSS is solving the imbalance problem.

In BSS, the imbalance problem means some stations are empty (there are no available bikes for renting) while some stations are full (there are no empty docks for returning). A customer will experience an unsatisfactory service when there are no available bikes or docks at a station. Improving customer satisfaction is one of the tasks for BSS operators.

In order to deal with the imbalance problem, the operators of BSS need to relocate bikes among stations by trucks. Some static truck-based strategies for relocating bikes have been proposed [4–6]. The repositioning time is usually at the system's idle time (such as midnight) in these truck-based strategies. Recently, some work has paid attention to dynamic truck-based strategies which relocate bikes based on the advanced prediction of bikes [7–9]. However, the truck-based approaches are not only expensive to operate but also violate the low-carbon goal of bike-share due to the amount of pollution from the trucks.

There are also some strategies that encourage customers to participate in bike allocation. For example, some bike-sharing systems incentivize customers to rent bikes from stations with many available bikes and return to those with many empty docks by giving fare discounts [10–12]. Implementing the customer-oriented rebalancing strategies should consider the following tasks:

(1) Determine the optimal number of bikes for each station; (2) Deal with the changing of bike demand over time in a day; (3) Suggest the incentive station which cannot be too far from customer's original target station; and (4) Set an appropriate bonus for the customer under a limited budget. In existing studies, the incentive strategies will be implemented only when a station's number of bikes is not enough. However, the station is already in a bad state when its number of bikes is not enough, which will effect the performance of incentive measures.

In this paper, we propose a strategy which can help the system choose a suitable time and determine a suitable station for implementing the incentive measure. Based on the proposed strategy, a bike-sharing system can effectively lengthen its uptime. First, the proposed strategy calculates the most suitable number of available bikes for each station. Then it determines the station which has the largest capacity for renting. In addition, the walking distance is also considered in the strategy. Take an instance in Figure 1—a customer intends to rent a bike at station S1. At present, S1 has eight available bikes, which is less than the optimal state (ten available bikes). According to the idea of our proposed strategy, a returning action is helpful for S1 to reach its optimal state rather than a renting action. Therefore, the system should incentivize him to rent a bike at station S2. S2 is within r1 meters away from S1. After this rebalancing, the states of S1 and S3 will not be worse. At the same time, S2 can also reach its optimal state. For the bike-sharing system, the result reduces the probability of the imbalance problem occurring in future. The main contributions of this work are listed as follows.

- An algorithm based on the Random Walk Process with two absorption walls (RWTAW) is proposed to calculate the optimal state of each station at a specific time. We prove that a station in the optimal state has the lowest possibility of being empty or full.
- An algorithm called the largest the first (TLTF) is proposed to choose the optimal station among the candidate stations on consideration of the walking distance and the minimal influence for the chosen station.
- Through establishing a simulation model of BSS based on historical datasets, we verify the effectiveness of our strategy for solving the imbalance problem compared with the truck-based static and dynamic strategies.

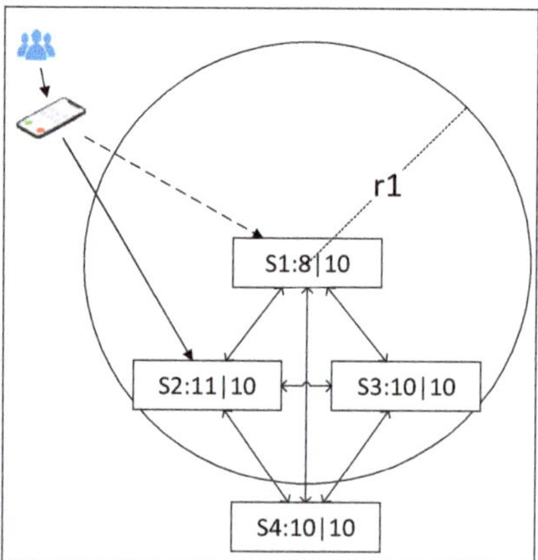

Figure 1. Example of renting process.

The rest of this paper is organized as follows. Section 2 gives an outline of the development of BSS and research on rebalancing optimization; the problem statement and the framework are presented in Section 3; Section 4 introduces the details of the RWTAW and TLTF algorithms; we talk about the performance of the proposed rebalancing strategy in Section 5; finally, we conclude this work and look forward to future research in Section 6.

2. Related Work

BSS has achieved great development, with the goal of solving the "last kilometer" problem in cities [13], since it was first built in Amsterdam in 1965. The first kind of BSS is the traditional dock-based bike sharing system in which a customer must rent and return the bike at a dock and the number of docks at each station is fixed, such as the CitiBike in New York and the Hubway in Boston. The second is the dockless bike-sharing system such as the Mobike and Ofo in China. Each bike has a unique QR code and GPS tracking model. These two technologies increase the ease of use and management. The biggest advantage of the dockless bike-sharing system is customers can easily rent bikes via a smartphone app and park the bikes at any valid places [14–16]. The convenience also brings some problems, for example, operators have to rebalance the bikes. The existing studies on BSS mainly focus on the following aspects.

2.1. System Rebalancing

In BSS, system rebalancing is the main method for solving the imbalance problem. The truck-based rebalancing strategies and the customer-oriented rebalancing strategies are two popular approaches. In this section, we will introduce these two approaches in detail.

2.1.1. Truck-Based Rebalancing

The truck-based rebalancing approaches have two main challenges, that is, determining the optimal inventory for each station and designing the optimal truck route with budget constraint. The approaches can be summarized into two categories according to the time of rebalancing. (1) Static Repositioning. Chemla et al. [4] used a branch-and-cut algorithm based on the tabu search to obtain the upper bounds and feasible routing solutions. However, their work only focused on solving the repositioning with a single vehicle in part of the city area. For the static bicycle repositioning problem under multiple vehicles, Raviv et al. [17] used the mixed integer linear program (MILP) to minimize the routing cost and customer dissatisfaction. Schuijbroek et al. [18] presented a rigorous cluster-first route-second heuristic to solve the inventory determining and routing decision problems. For the dockless system, Liu et al. [19] developed an enhanced version of chemical reaction optimization to solve the static bike repositioning problem by considering multiple heterogeneous vehicles, multiple depots and multiple visits. Pal and Zhang [5] presented a Novel Mixed Integer Linear Program for solving the static rebalancing problem in a free-floating bike sharing system. They also presented a hybrid nested large neighborhood search for large-scale bike sharing programs. (2) Dynamic Repositioning. Chiariotti et al. [20] used the Birth-Death Process to model a station's occupancy and proposed a dynamic rebalancing strategy. They first determined the optimal inventory for stations and then modeled the path-planning problem as a vehicle routing problem and adopted the greedy heuristic to dynamically redistribute bikes every hour. Contardo et al. [8] presented a mathematical programming formulation for the dynamic repositioning problem and obtained the lower bounds and feasible solutions by using the decomposition schemes. Li et al. [9] proposed a spatio-temporal reinforcement learning (STRL) model for dynamic bike repositioning. They first proposed an inter-independent inner-balance clustering algorithm to cluster stations into groups. Then they utilized the O-Model and the I-Model to predict the demands of renting and returning respectively. Lastly, for each cluster, the STRL obtains the optimal inner-cluster reposition policy based on a deep neural network. Liu et al. [21] developed a multi-source optimization approach for the rebalancing problem. They proposed an Adaptive Capacity Constrained K-centers Clustering

(AdaCCKC) algorithm which can reduce the large-scale multiple vehicle routing problem to an inner cluster one vehicle routing problem with guaranteed feasible solutions. Then the vehicle rebalancing route can be optimized by a mixed integer nonlinear programming (MINLP) model.

The drawbacks of the static and dynamic truck-based strategies are the expensive truck operating costs and the pollution of tail gas. Some researchers therefore now pay more attention to the customer-oriented rebalancing strategies.

2.1.2. Customer-Oriented Rebalancing

The customer-oriented rebalancing approach is to solve the imbalance problem by providing customer incentives (bonus, fare discount or otherwise) without deploying trucks. For example, Singla et al. [11] proposed a method on the basis of a crowdsourcing mechanism [22] to incentivize customers to rent or return bikes at specific station. In their work, the dynamic pricing mechanism with a limited budget was proposed. It utilizes the regret minimization approach to maximize rebalancing efficiency. Chemla et al. [10] proved that the dynamic regulation problem is NP-hard (non-deterministic polynomial-time) and presented a heuristic to circumvent it. In their work, the pricing technique based on the linear programming was proposed. Fernandez et al. [23] designed a bike sharing system simulator to evaluate incentive-based rebalancing strategies. Some incentive strategies have been adopted in BSS. For instance, the work of Fricker and Gast [12] was used in the Velib+ system (in Paris). The incentive strategy is that customers will receive additional riding time for free if they return bikes at uphill stations. They also proved that the system can improve the operating efficiency if the customers who enjoy the incentive strategy park their bikes at two or more stations in a bad state. According to the current and predicted state of the system, Pfrommer et al. [24] proposed the price incentive scheme which takes into account the model-based predictive control principles. In a free-floating bike sharing system, Pan et al. [25] proposed a novel deep reinforcement learning framework for incentivizing users. Their work modeled the rebalancing problem as a Markov decision process and took both spatial and temporal features into consideration. The proposed Hierarchical Reinforcement Pricing (HRP) algorithm shows great performance on a dataset from Mobike, a major Chinese dockless bike sharing system. More recently, Diez et al. [26] proposed a persuasion model to recommend the optimal station and the shortest riding route between two locations. The model combines the argumentation theory and the characteristics of customers. In our work, the rebalancing strategy, which is based on the Random Walk process and integrates the walking distance between stations, was designed. The proposed strategy helps to lengthen the station's working time by inferring the optimal station for customers.

2.2. System Prediction

The prediction of bike demand is crucial for the bike repositioning problem. An accurate prediction can improve the effect of repositioning. Some station-level predicting models have been proposed. For example, Zeng et al. [27] proposed a station-centric model which takes into account the global features such as weather, user activity and season. Kaltenbrunner et al. [28] adopted the auto-regressive moving-average (ARMA) model to predict the number of bikes and docks for each station. Similarly, Yoon et al. [29] used a modified autoregressive integrated moving average (ARIMA) model to predict the available resources at each station. In their work, the spatial interaction and temporal factors are considered. Unlike the station-level prediction, some models are proposed according to the clustering of stations with similar patterns of usage. For example, Li et al. [30] proposed a hierarchical prediction model to predict the check-out/in of each station cluster. They first clustered the stations by the bipartite clustering algorithm based on the geographical locations of stations and transition patterns and then predicted the entire traffic in the city by way of the gradient boosting regression tree (GBRT) model. Finally, the rental proportion across clusters was predicted by a multi-similarity-based inference model. Chen et al. [31] proposed a dynamic cluster-based framework for over-demand bike prediction. Recently, neural networks have been adopted to predict bike demand. Liu et al. [32] proposed a prediction model based on artificial neural networks (ANN). The model utilizes a set of features such as human mobility data, POI and station network structures. Lin et al. [33] proposed a novel graph convolutional neural network to predict the hourly demand in a BSS.

2.3. System Design and Traffic Pattern Analyzing

System design, including the station layout, capacity, and price policy, is the premise of successfully operating a bike sharing system. Dell'Olio et al. [34] calculated the potential demand of bike usage in a city and the price the customer would like to pay. They also proposed a location model with the help of geographical information about the stations. Chen and Sun [35] proposed a mathematical model to formulate the layout of public bike stations with the objective of minimizing users' total travel time. Lin and Yang [36] designed the strategic planning of BSS with service level considerations. They proposed an optimization model including a nonlinear integral and greedy heuristic to help operators determine the number of stations, the network structure of bike paths connecting the stations and the travel paths for users between each origin and destination station. In addition to the system design and efficient operation, the traffic pattern analysis is another important topic, such as the bike usage pattern and the transition pattern across stations in BSS [37–39]. Understanding the traffic pattern of BSS is helpful for operating the system efficiently and knowing the mobility of a city. Furthermore, the massive real trajectories generated by users help to solve some city problems, such as bike lanes planning [40] and illegal parking detection [41].

3. Design Overview

3.1. Problem Statement

In this paper, we focus on a bike sharing system in which the station is fixed. Generally, a BSS has n stations. The set of stations is defined as $S = \{S_1, S_2, \ldots, S_n\}$. The maximum capacity of station S_i is defined as M_i. $m_i(t)$ represents the available bikes at station S_i at time t. We model the bike sharing system as a fully connected graph $G = \{S, I\}$. Each node $I_{i,j} = (S_i, S_j)$ corresponds to the edge between two stations S_i and S_j. The edges are weighed by walking distance metric $r_{i,j}$. Notations used in this paper are shown in Table 1.

We set the scene of rebalancing as follows. A customer will query the information about stations on his mobile phone before his trip. After the customer selects the station at which he will rent a bike, the bike system immediately checks the station's state. If the number of the station's available bikes is less than its optimal state, the system would incentivize the customer to another station according to the states of nearby stations and the walking distance. We hope the whole system can obtain the least values of three metrics, that is, the unworking time, the proportion of lost customers and the times of reposition. The detailed definitions of the three metrics are introduced in Section 5. The selection of the recommended station at which the system incentivizes the customer to rent bikes is the main task in this work. Formally, the problem is defined as follows:

Problem Definition: Given the station S_i at which a customer will rent a bike, the bike system will determine whether to incentivize the customer to another station S_j or not. If needed, the bike system should output the recommendation station S_j.

Table 1. Notations.

n	Total number of stations
$I_{i,j}$	The edge between station S_i and S_j
M_i	Capacity of station S_i
$m_i(t)$	Available bikes of station S_i at time t
$r_{i,j}$	Walking distance from S_i to S_j
$p_{i,j}$	The probability ride from S_i to S_j
$T_{i,j}$	Trip duration by bike from S_i to S_j
$t(k)$	The time period in the day indexed by slice k
$\lambda_i(k)/\mu_i(k)$	The customer arrival rate for renting/returning at S_i
$OP_{i,k}$	The optimal state of S_i during time $t(k)$
OS_t	The optimal station for renting at time t

3.2. Framework

Figure 2 shows the framework of our bike-sharing system with the rebalancing strategy, which consists of the following three components:

- Customer Module: This module is designed to simulate the behaviors of customers. In BSS, there are two processes that need to be modeled. One is the customers' arrival process for renting at a station and the other is the arrival process for returning. Previous studies [42,43] have demonstrated that these two processes can be described by the inhomogeneous Poisson process. According to the definition of the Poisson process, in BSS, we need to determine the arrival rate. $\lambda_i(t)$ and $\mu_i(t)$ are defined as the arrival rates corresponding to the two processes, respectively. The arrival rate represents the number of customers arriving at the station per unit time. Therefore, $\lambda_i(t)$ and $\mu_i(t)$ are two important parameters in the customer module. Moreover, the trip duration is necessary for each trip. According to the analysis of historical data (as shown in Section 5), the distribution of trip duration fits well with the lognormal distribution. Therefore, the trip duration for a certain trip is a random number of log-normal distribution in our simulated BSS.
- Station Module: We design this module to simulate the states of stations in the BSS. Each station in the system has three properties: geographic location, capacity and transition matrix. The geographic location is decided by the latitude and longitude. The capacity represents the number of docks. The transition matrix includes the probabilities that customers ride bikes from station S_i (departure station) to other stations. The geographic location and capacity are defined in the system. The transition matrix can be obtained by the historical records.
- System Control Module: This module is the core part of rebalancing in the BSS. The optimal state of a station is calculated by the RWTAW algorithm. The TLTF algorithm calculates the optimal station based on the walking distance and the current states of the candidate stations. The control module is the main contribution of this work. We will describe the details of the RWTAW and TLTF algorithms in the next section.

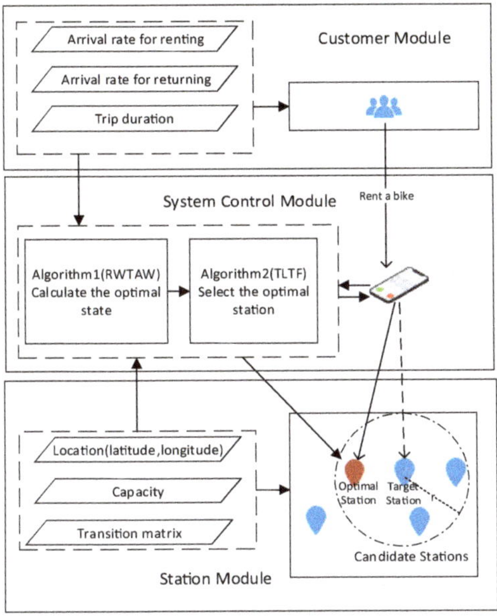

Figure 2. Framework of simulate system.

4. Methodology

4.1. The Optimal State Calculating

In the BSS, we hope the station can keep the optimal state as long as possible, thus it should have the lowest possibility that it will be empty or full in future. Motivated by this idea, $OP_{i,k}$ is defined as the optimal state of S_i at time slice k. It can be calculated on the basis of the probability that station becomes either empty state ($m_i(t) = 0$) or full state ($m_i(t) = M_i$). Formally, it can be defined as follows:

$$OP_{i,k} = \arg\min_{a} P_a, \forall a \in \{1, 2, \ldots, M_i - 1\} \tag{1}$$

where P_a is the probability that station becomes either empty or full during time $t(k)$ and it is calculated by:

$$P_a = P_{a \to 0} + P_{a \to M_i} \tag{2}$$

where $P_{a \to 0}$ and $P_{a \to M_i}$ respectively represent the probabilities that station S_i becomes empty state and full state when the current state $m_i(t) = a$.

Random Walk is a mathematical statistical model [44,45], which has a wide range of applications. For example, researchers have studied the application of the Random Walk model in finance prediction [46], high level data classification [47] and social network optimization [48,49]. A one-dimensional random walk can be looked at as a Markov chain whose state space is given by the integers τ ($\tau = 0, \pm 1, \pm 2, \ldots$). The Random walk with two absorption states (τ_1 and τ_2) at both ends is an important model. The model stops walking once it reaches either absorption state. For an initial state τ_a ($\tau_1 < \tau_a < \tau_2$), the probability P_{τ_a} represents walking from τ_a to either absorption state. P_{τ_a} can be calculated as follows:

$$P_{\tau_a} = \frac{(\frac{q}{p})^{\tau_a - \tau_1} - (\frac{q}{p})^{\tau_2}}{1 - (\frac{q}{p})^{\tau_2}} \tag{3}$$

where p represents the probability of walking toward τ_2 at each step, q represents the probability of walking toward τ_1 at each step.

In this paper, we describe the state changing process of each station by the one-dimensional Random Walk process. In a bike sharing system, a station's state has only two changing directions: the empty state ($m_i(t) = 0$) and the full state ($m_i(t) = M_i$). The state of a station goes one step toward the empty state when a renting event occurs. In the same way, the state of station goes one step toward the full state when a returning event occurs. For a middle state $m_i(t) = a$, if $0 < a < OP_{i,k}$, the smaller the value of a, the larger the probability that the station will become empty; if $OP_{i,k} < a < M_i$, the larger the value of a, the larger the probability that the station will become full. We assume that the station will stop working and wait for supplementary as long as its state becomes either empty or full. Thus, the empty state and full state of a station correspond to two absorption walls.

Specifically, given the total changing steps e_1 and e_2, if the station becomes empty after e_1 steps, the probability for this case is calculated by the following formula:

$$P_{a\to 0} = \sum_{i=0}^{e_1} C^i_{a+2i} \times P^{a+i}_{rent} \times P^i_{return} \tag{4}$$

Similarly, $P_{a\to M_i}$ is given as:

$$P_{a\to M_i} = \sum_{i=0}^{e_2} C^i_{M_i-a+2i} \times P^{M_i-a+i}_{return} \times P^i_{rent} \tag{5}$$

P_{rent} and P_{return}, respectively, represent the probabilities of a renting event and a returning event for each step at a station. The constraints of total steps e_1 and e_2 are:

$$e_1 = \min\{Rent_{i,k} - a, Return_{i,k}\} \tag{6}$$

$$e_2 = \min\{a - M_i + Return_{i,k}, Rent_{i,k}\} \tag{7}$$

In a bike sharing system, the process by which customers arrive at a station to rent or return bikes can be described by the Poisson process [43]. The renting intensity of Poisson process $Rent_{i,k}$ corresponds to the total number of customers who rent at station S_i during time $t(k)$. The returning intensity of Poisson process $Return_{i,k}$ corresponds to the total number of returning customers. $Rent_{i,k}$ and $Return_{i,k}$ can be calculated as follows:

$$Rent_{i,k} = \lambda_i(k) \times T \tag{8}$$

$$Return_{i,k} = \mu_i(k) \times T \tag{9}$$

where $\lambda_i(k)$ and $\mu_i(k)$ correspond to the customer arrival rate for renting and returning respectively. T is the length of time slice k. It is worth noting that we divide a day into 24 time slices. For instance, $\lambda_i(8)$ means the arrival rate for renting from 7:00 am to 8:00 am.

According to the $Rent_{i,k}$ and $Return_{i,k}$, the probabilities of renting (P_{rent}) and returning (P_{return}) at each step at time slice k are defined respectively as follows:

$$P_{rent} = \frac{Rent_{i,k}}{Rent_{i,k} + Return_{i,k}} \tag{10}$$

$$P_{return} = \frac{Return_{i,k}}{Rent_{i,k} + Return_{i,k}} \tag{11}$$

In Equations (10) and (11), we ignore the influence of customers' arrival order for renting and returning and only focus on the total number of customers that have arrived in this time period [20].

The above discussions indicate that the proposed RWTAW algorithm can describe the changing state of a station during a certain period. Algorithm 1 gives the details of calculating $OP_{i,k}$. This proposed algorithm can be computed in $\mathcal{O}(n)$ operations. The total runtime of the algorithm is determined by the capacity of the station. Next, we will introduce how to choose the optimal station at which the system incentivizes the customer to rent bikes.

Algorithm 1: RATAW

Input: $\lambda_i(k)$, $\mu_i(k)$: customer arrival rate for renting and returning, M_i: capacity of station, T: length of time slice
Output: $OP_{i,k}$: optimal state

for *station* $S_i \in S$ **do**
  for *time slice* $k \in \{1, 2, \ldots, 24\}$ **do**
    init $a = 0, P = P_0, OP_{i,k} = 0$;
    while $a \leq (M_i - 1)$ **do**
      $a = a + 1$;
      compute $P_{a \to 0}$ and $P_{a \to M_i}$ by Equation (4), Equation (5) ;
      compute P_a by Equation (2) ;
      if $P_a < P$ **then**
        $P = P_a$;
        $OP_{i,k} = a$;
      end
    end
  end
end
Return $OP_{i,k}$;

4.2. The Optimal Station Selecting

When a customer chooses his target station S_i at time t on the APP (application program) of BSS, the system would incentivize him to rent bikes at another optimal station OS_t if S_i is not in its optimal state ($m_i(t) < OP_{i,k}$). Two issues need to be solved before deciding the optimal station: (1) The selected station should be as close as possible to the customer's target station, because customer may be unwilling to walk too far; and (2) The incentive measure should have a minimal impact on the station's sustainable working or can help the station be closer to its optimal state. Aiming to solve these two issues, we proposed the TLTF algorithm to determine the optimal station OS_t which satisfies:

$$OS_t = \arg\max_{S_j} \delta_{j,k}, \forall S_j \in C_i \tag{12}$$

where C_i is the set of candidate stations. We choose C_i based on the maximum distance r that customers are willing to walk [11]. Specifically, we select the stations which are no more than r from S_i as the candidate stations in C_i:

$$C_i = \{S_j \mid d(S_i, S_j) \leq r\} \tag{13}$$

In Equation (12), $\delta_{i,k}$ represents the difference between the current state of station S_i and its optimal state at time slice k, which is the key influence on OS_t. Its definition is given as follows:

$$\delta_{i,k} = m_i(t) - OP_{i,k} \tag{14}$$

If $\delta_{i,k} > 0$, the rebalancing strategy will not be triggered when a customer chooses S_i for renting. If $\delta_{i,k} \leqslant 0$, the system will incentivize him to rent bikes at station OS_t. As defined in Equation (12), the station OS_t is the one which has the largest difference between its current state and the optimal state. The procedure to determine the optimal station OS_t is described in Algorithm 2. The complexity of the algorithm is $\mathcal{O}(n)$.

Algorithm 2: TLTF

Input: $OP_{i,k}$:optimal state for S_i at time slice k, $D_{1\times n-1}$:matrix of distance between S_i and left $n-1$ stations, r: distance

Output: OS_t:optimal station the system incentivizes customer to rent a bike

Note that $D_{1\times n-1} = \{d(S_i,S_1), d(S_i,S_2), \ldots, d(S_i,S_{n-1})\}$;
init C_i: set of candidate stations ;
foreach *station* $S_j, j = 1,2,\ldots,i-1,i+1,\ldots,n$ **do**
    if $d(S_i,S_j) \leq r$ **then**
        add S_j into C_i ;
    end
end
compute $\delta_{i,k}$ by Equation (14) ;
if $\delta_{i,k} > 0$ **then**
    the customer rent at station S_i ;
else
    compute δ for stations in C_i ;
    determine the optimal station OS_t by Equation (12) ;
    incentivize customer to rent at station OS_t ;
end

---

5. Experimental

5.1. Data Set Description

The datasets used in this paper were obtained from Bay Area, a bike sharing system in San Francisco. We have uploaded the datasets to github (https://github.com/TwinkleBill/babs_open_data_year_3). The datasets include 68 stations' records from 1 September 2015 to 31 August 2016. Each record includes the status of station and customer's trajectory information. We give a visualization of the stations in Figure 3 and show the statistical information of the datasets in Table 2.

Figure 3. Map of the Bay Area bike sharing system. (**a**) Stations in Bay Area; (**b**) Stations in our simulation.

Table 2. Details of the Dataset.

Data source	Bay Area
Time span	1 September 2015 to 31 August 2016
Stations	68 (ID, name, latitude, longitude, capacity, city, installation date)
Trajectory trips	314,000
Status records	36 million (the number of available bikes and docks)

5.2. Performance Metrics for Bike-Sharing System

Here we define three metrics to evaluate the performance of the rebalancing strategies in a bike sharing system.

- The unworking time: The sum of the duration time of the stations which keep empty or full. The unsatisfactory service increases when a station is in its unworking time, since customers have no available bikes or docks to use. The smaller the unworking time of the system, the smaller the probability of failing to rent bikes or return bikes.
- The proportion of lost customers (γ_{lost}): γ_{lost} is defined as follows:

$$\gamma_{lost} = \frac{n_{lost}}{n_{lost} + n_{ride}} \tag{15}$$

where n_{lost} is the number of customers who fail to rent or return a bike, n_{ride} is the number of customers who successfully ride a bike. The smaller the value of γ_{lost}, the better the performance of the system.
- The times of reposition: The number of bikes that need to be rebalanced (the truck-based rebalancing strategies) or the number of customers (the customer-oriented rebalancing strategies). In terms of the truck-based rebalancing strategies, the times of relocating bikes corresponds to the truck operating costs. For customer-oriented rebalancing strategies, the incentive cost also increases with the increasing number of incentive customers. To simplify things, the cost of delivering a bike is seen as the same as incentivizing a customer. Obviously, the smaller the times of reposition, the less the costs of rebalancing.

5.3. Parameter Setting

In the simulative bike sharing system, we need to set the following parameters: (1) λ_i and μ_i, the customer arrival rates for renting and returning at station S_i respectively. (2) The distribution parameters of trip duration $T_{i,j}$. (3) The walking distance r in the TLTF algorithm.

As discussed before, for each station in a BSS, the arrival processes of customers for renting and returning can be described by the Poisson process [43], which indicates that the time interval between two adjacent events should obey the exponential distribution. We analyzed the time interval distribution of the experimental data and found that most of the stations followed the exponential distribution. We take the historical records of the station named "San Jose Diridon Caltrain Station" from 7:00 a.m. to 8:00 a.m. (the morning peak [50], as shown in Figure 4) as an example. Figure 5a shows the time interval distribution of the renting process. It follows the exponential distribution with the parameter $\lambda = \dfrac{1}{11.5208}$. Thus, we utilize the Poisson process with $\lambda_i = \dfrac{1}{11.5208}$ to simulate the renting process of customers at this station from 7:00 a.m. to 8:00 a.m. Similarly, as shown in Figure 5b, the Poisson process with $\mu_i = \frac{1}{8.4724}$ is used to simulate the returning process. In the RWTAW algorithm, the values of λ_i and μ_i of each station are different.

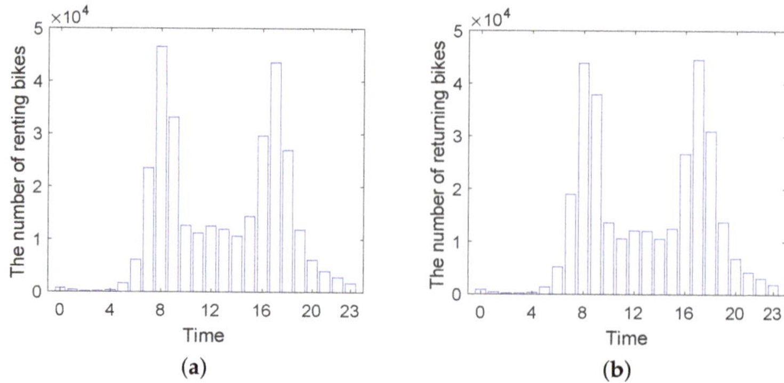

Figure 4. The renting and returning bikes by hours. (**a**) Renting by hours; (**b**) Returning by hours.

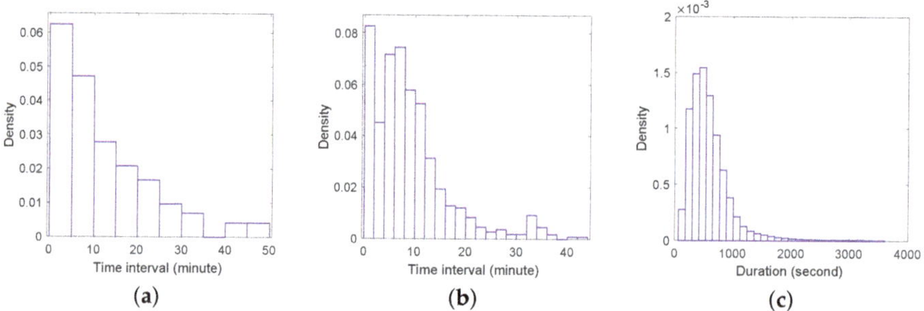

Figure 5. Statistical information of time interval and trip duration. (**a**) Rent; (**b**) Return; (**c**) Trip Duration.

The trip duration is also an important parameter in the simulation system. As shown in Figure 5c, the distribution of trip duration obeys the lognormal distribution (the mean is 6.2806 and the sigma is 0.7032). In our simulation system, the trip duration of each trip is generated by this lognormal distribution.

In order to demonstrate the effectiveness of our simulated bike sharing system, the result of one simulation of the station named "San Jose Diridon Caltrain Station" is illustrated in Figure 6. We plot the variation of the number of bikes rented by customers. The changing pattern of the simulation is in line with that of the empirical data. According to the results of the simulation, we have confidence that the simulated bike sharing system with the aforementioned parameters can reproduce the results of a real system.

In TLTF, the value of walking distance r is crucial, because it will take more incentivizing bonus to encourage customers to rent bikes at stations further away. Furthermore, the r is also related to the computational cost of the algorithm. In other words, a large value of r means lots of stations will be selected as candidate stations by the algorithm when it calculates the OS_t. On the other side, the increase of r may help to select the optimal station for renting. Figure 7 shows the influence of distance r on the performance of the system. When r increases from 600 m to 1000 m, the unworking time and the proportion of lost customers decrease. Because the candidate stations will include more stations with an increase of the walking distance r. For example, in the datasets, S_{14} is the only candidate of S_2 when $r = 600$ m, while the candidate stations consist of S_4, S_5 and S_{14} when $r = 1000$ m. However, the result will be bad if the walking distance is too far. As we can see, the unworking time and the proportion of lost customers increase when $r = 1200$ m. Although a large r increases the number of candidate stations, it also leads to an increase in the size of the intersection of candidate stations. These stations will be recommended by the system. In this case, they will become empty or full soon. So the unworking time and the proportion of lost customers increase when $r = 1200$ m. We also care about the relationship between the probability of a customer's acceptance of walking distance and people's travel behavior [51], and count each station's walking distance to its adjacent stations which are not more than 1000 m away. We find that the stations uniformly distribute within 1000 m away from a center station. We think the acceptance probability of this walking distance is also suitable for customers, because they have a wide choice for purpose station [11]. Therefore, we set $r = 1000$ m.

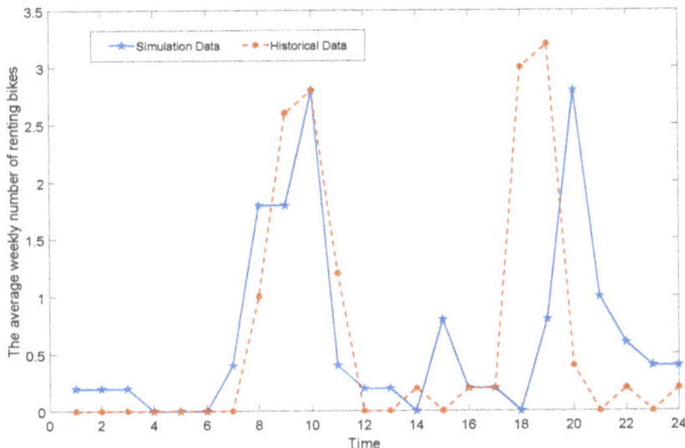

Figure 6. Simulation Results of station named "San Jose Diridon Caltrain Station".

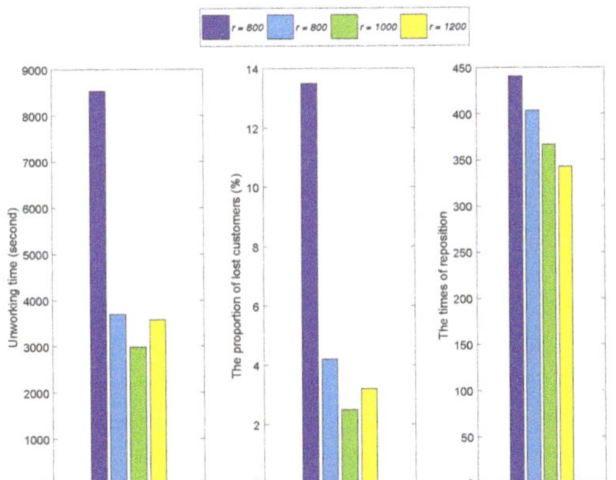

Figure 7. System performance with different r in TLTF.

In addition, we also analyze bike usage on workdays and weekends. As shown in Figure 8, the patterns of the system on workdays are similar, that is, two peaks respectively appear at 10:00 a.m. and 18:00 p.m. One is the morning peak, the other is the evening peak (the two peaks for whole system are 8:00 a.m. and 18:00 p.m., as shown in Figure 4a). However, the patterns on weekends are distinctly different. On weekends, the usage is lower than that on workdays, and is irregular. Such a phenomenon is consistent with what we have mentioned before, that is, commuting time is an important factor which leads to the stable traffic pattern on workdays. Therefore, in this work, we only simulate the operation of the bike sharing system on workdays. The parameters of the system are the same on each workday.

Note that in the dataset of Bay Area, stations are distributed into four districts (as shown in Figure 3a). There are no trajectories between any two different districts. The reason is that customers generally do not choose bikes for long-distance travel. Therefore, stations in our simulation system are all in "San Jose" which has 15 stations (as shown in Figure 3b). The probability of choosing the returning station for the customer satisfies the uniform distribution. Experiments indicate that such an assumption does not influence the time interval distribution of a customer's arrival process for returning to a station, which still fits well with the exponential distribution.

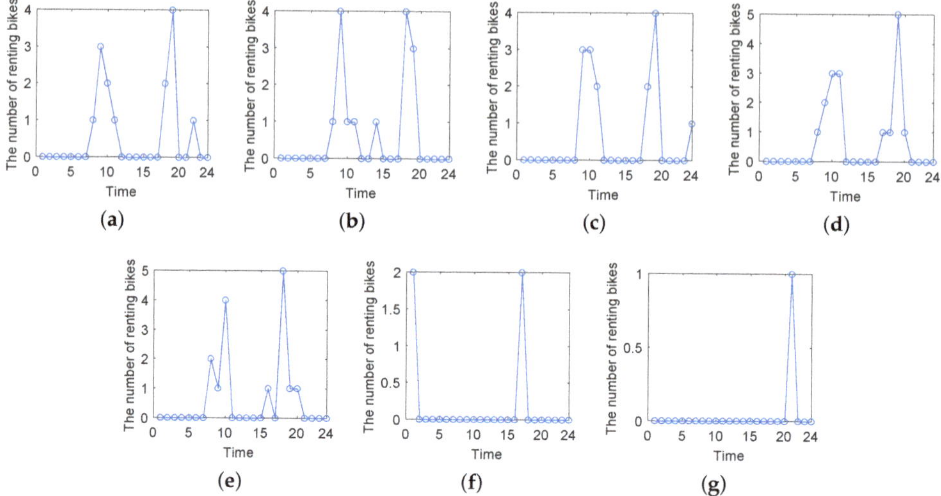

Figure 8. Bike usage on workdays and weekends. (**a**) Monday; (**b**) Tuesday; (**c**) Wednesday; (**d**) Thursday; (**e**) Friday; (**f**) Saturday; (**g**) Sunday.

5.4. Results

In order to validate the proposed strategy, we compared it with three scenarios: the system with no rebalancing, the static rebalancing strategy and the dynamic rebalancing strategy. Note that our target is to verify the effectiveness of the rebalancing strategy, so we do not take too much care on the truck routing problem and the price strategy of incentive in this paper.

- No Rebalancing (NR): We operate the system without any intervention, even if there appears full or empty stations.
- Static Rebalancing (SR): The static rebalancing strategy is that we bring the stations to their optimal state once a day, at 3:00 a.m., as in Reference [7].
- Dynamic Rebalancing (DR): The dynamic rebalancing strategy proposed in Reference [20] calculates the optimal state for stations first, and then makes each station be in its optimal state at the beginning of every i hours. In our experiments, we simulate three DR strategies, that is, $i = 1, 4, 8$.

Figure 9 shows the system's performance with different rebalancing strategies. The system with no rebalancing gets the worst performance on the unworking time and the proportion of lost customers. Compared with the NR strategy, the SR strategy decreases by nearly 40% and 39% on these two metrics, respectively. With the decrease of the scheduling time interval, the effectiveness of the dynamic strategy improves (in $DR_i, i = 1, 4, 8$). As shown in Figure 9, among the three DR strategies, the one which has the shortest scheduling time interval ($i = 1$) has the least unworking time and the smallest proportion of lost customers. However, its times of reposition is the largest. According to the simulation results, the proposed rebalancing strategy can obtain the smallest proportion of lost customers with the proper times of reposition. At the same time, the unworking time is also within our acceptance.

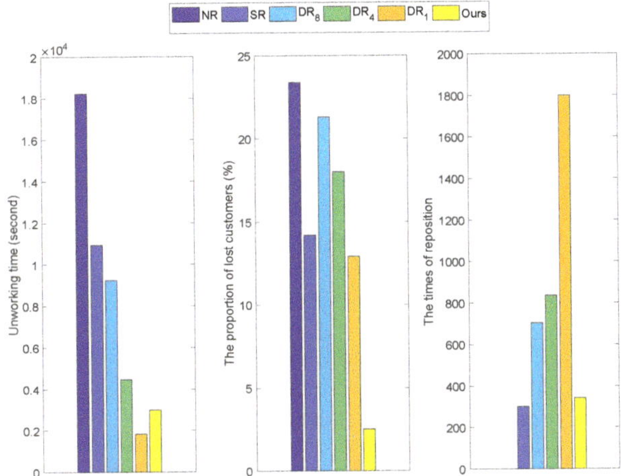

Figure 9. Comparison of system performance.

In BSS, the traffic pattern changes rapidly. The rapid response of the system is important for offering customers a better service. Thus, the rebalancing algorithm should be in real-time. For static strategies, they are implemented at the system's idle time, which means there is no user in the system during this period. So we do not have to pay attention to the real-time performance of static strategies. Table 3 shows the runtime of our strategy and the DR strategy [20] to calculate a station's optimal state at a specific hour. There are four stations. The capacity of each station is 11, 15, 19 and 27 respectively. As we can see, the increase of the station's capacity increases the computing time. The time cost of the DR strategy is greater than ours. The results show that our strategy is real-time and efficient.

Table 3. Comparison of runtime of calculating the optimal state.

StationID (Capacity)	DR		Ours	
	Runtime (s)	Optimal State	Runtime (s)	Optimal State
2(27)	13.352	8	0.026	6
5(19)	9.765	5	0.008	5
6(15)	10.533	8	0.014	9
4(11)	7.224	8	0.012	7

6. Conclusions

In this paper, we designed a customer-oriented rebalancing strategy for bike sharing systems. On the basis of the one-dimension Random Walk process with two absorption walls, we infer the probability that a station becomes empty or full, which is useful to both system operators and customers. The proposed strategy includes two algorithms called RWTAW and TLTF respectively. The RWTAW calculates the optimal state of the station. The TLTF determines the optimal station at which the system encourages customers to rent. Compared with the truck-based static and dynamic rebalancing methods, our strategy has the best performance on decreasing the imbalance of the system while keeping the rebalancing cost as low as possible. In future work, we will explore how to predict the number of bikes for each station to improve customer satisfaction.

Author Contributions: conceptualization, P.Y. and F.H.; methodology, P.Y.; software, P.Y.; validation, P.Y., F.H. and J.P.; formal analysis, P.Y.; investigation, F.H.; resources, J.P.; data curation, P.Y.; writing–original draft preparation, P.Y. and F.H.; writing–review and editing, P.Y., F.H. and J.P.; visualization, P.Y.; supervision, P.Y.; project administration, J.P.; funding acquisition, J.P.

Funding: This research was funded by the National Key Research and Development Project with Grant No. 2017YFB0202403, Sichuan Key Research and Development Program (2017GZDZX0003-02, 2019KJT0015-2018GZ0098), and the Sichuan University-Zigong Science and Technology Fund (2018CDZG-15).

Conflicts of Interest: The authors declare no conflict of interest.

References

1. Lin, B.; Zhu, J. Impact of energy saving and emission reduction policy on urban sustainable development: Empirical evidence from China. *Appl. Energy* **2019**, *239*, 12–22. [CrossRef]
2. Fishman, E.; Washington, S.; Haworth, N. Bike share's impact on car use: Evidence from the United States, Great Britain, and Australia. *Transport. Res. Part D-Transport. Environ.* **2014**, *31*, 13–20. [CrossRef]
3. Zhang, Y.; Mi, Z. Environmental benefits of bike sharing: A big data-based analysis. *Appl. Energy* **2018**, *220*, 296–301. [CrossRef]
4. Chemla, D.; Meunier, F.; Calvo, R.W. Bike sharing systems: Solving the static rebalancing problem. *Discret. Optim.* **2013**, *10*, 120–146. [CrossRef]
5. Pal, A.; Zhang, Y. Free-floating bike sharing: Solving real-life large-scale static rebalancing problems. *Transp. Res. Part C-Emerg. Technol.* **2017**, *80*, 92–116. [CrossRef]
6. Erdoğan, G.; Battarra, M.; Calvo, R.W. An exact algorithm for the static rebalancing problem arising in bicycle sharing systems. *Eur. J. Oper. Res.* **2015**, *245*, 667–679. [CrossRef]
7. O'Mahony, E.; Shmoys, D.B. Data Analysis and Optimization for (Citi)Bike Sharing. In Proceedings of the Twenty-Ninth AAAI Conference on Artificial Intelligence, AAAI'15, Austin, TX, USA, 25–30 January 2015; AAAI Press: Menlo Park, CA, USA, 2015; pp. 687–694.
8. Contardo, C.; Morency, C.; Rousseau, L.M. *Balancing a Dynamic Public Bike-Sharing System*; Technical Report; Cirrelt: Montreal, QC, Canada, 2012.
9. Li, Y.; Zheng, Y.; Yang, Q. Dynamic Bike Reposition: A Spatio-Temporal Reinforcement Learning Approach. In Proceedings of the 24th ACM SIGKDD International Conference on Knowledge Discovery & Data Mining, London, UK, 19–23 August 2018; ACM: New York, NY, USA, 2018; pp. 1724–1733.
10. Chemla, D.; Meunier, F.; Pradeau, T.; Calvo, R.W.; Yahiaoui, H. *Self-Service Bike Sharing Systems: Simulation, Repositioning, Pricing*; Technical Report; Centre d'Enseignement et de Recherche en Mathematiques et Calcul Scientifique (CERMICS): Marne la Vallée, France, 2013.
11. Singla, A.; Santoni, M.; Bartók, G.; Mukerji, P.; Meenen, M.; Krause, A. Incentivizing Users for Balancing Bike Sharing Systems. In Proceedings of the Twenty-Ninth AAAI Conference on Artificial Intelligence, Austin, TX, USA, 25–30 January 2015; AAAI Press: Menlo Park, CA, USA, 2015, pp. 723–729.
12. Fricker, C.; Gast, N. Incentives and redistribution in homogeneous bike-sharing systems with stations of finite capacity. *EURO J. Transp. Logist.* **2014**, *5*, 261–291. [CrossRef]
13. DeMaio, P. Bike-sharing: History, Impacts, Models of Provision, and Future. *J. Publ. Transp.* **2009**, *12*, 41–56. [CrossRef]
14. Wang, M.; Zhou, X. Bike-sharing systems and congestion: Evidence from US cities. *J. Transp. Geogr.* **2017**, *65*, 147–154. [CrossRef]
15. Fishman, E. Bikeshare: A Review of Recent Literature. *Transp. Rev.* **2015**, *36*, 92–113. [CrossRef]
16. Shen, Y.; Zhang, X.; Zhao, J. Understanding the usage of dockless bike sharing in Singapore. *Int. J. Sustain. Transp.* **2018**, *12*, 686–700. [CrossRef]
17. Raviv, T.; Tzur, M.; Forma, I.A. Static repositioning in a bike-sharing system: models and solution approaches. *EURO J. Transp. Logist.* **2013**, *2*, 187–229. [CrossRef]
18. Schuijbroek, J.; Hampshire, R.; van Hoeve, W.J. Inventory rebalancing and vehicle routing in bike sharing systems. *Eur. J. Oper. Res.* **2017**, *257*, 992–1004. [CrossRef]
19. Liu, Y.; Szeto, W.; Ho, S.C. A static free-floating bike repositioning problem with multiple heterogeneous vehicles, multiple depots, and multiple visits. *Transp. Res. Pt. C-Emerg. Technol.* **2018**, *92*, 208–242. [CrossRef]
20. Chiariotti, F.; Pielli, C.; Zanella, A.; Zorzi, M. A Dynamic Approach to Rebalancing Bike-Sharing Systems. *Sensors* **2018**, *18*, 512. [CrossRef] [PubMed]

21. Liu, J.; Sun, L.; Chen, W.; Xiong, H. Rebalancing bike sharing systems: A multi-source data smart optimization. In Proceedings of the 22nd ACM SIGKDD International Conference on Knowledge Discovery and Data Mining, San Francisco, CA, USA, 13–17 August 2016; ACM: New York, NY, USA, 2016; pp. 1005–1014.
22. Sadilek, A.; Krumm, J.; Horvitz, E. Crowdphysics: Planned and opportunistic crowdsourcing for physical tasks. In Proceedings of the Seventh International AAAI Conference on Weblogs and Social Media, Boston, MA, USA, 8–11 July 2013.
23. Fernández, A.; Timón, S.; Ruiz, C.; Cumplido, T.; Billhardt, H.; Dunkel, J. A Bike Sharing System Simulator. In Proceedings of the International Conference on Practical Applications of Agents and Multi-Agent Systems, Toledo, Spain, 20–22 June 2018; Springer: Berlin, Germany, 2018; pp. 428–440.
24. Pfrommer, J.; Warrington, J.; Schildbach, G.; Morari, M. Dynamic Vehicle Redistribution and Online Price Incentives in Shared Mobility Systems. *IEEE Trans. Intell. Transp. Syst.* **2014**, *15*, 1567–1578. [CrossRef]
25. Pan, L.; Cai, Q.; Fang, Z.; Tang, P.; Huang, L. A Deep Reinforcement Learning Framework for Rebalancing Dockless Bike Sharing Systems. *arXiv* **2018**, arXiv:1802.04592.
26. Diez, C.; Palanca, J.; Sanchez-Anguix, V.; Heras, S.; Giret, A.; Julián, V. Towards a Persuasive Recommender for Bike Sharing Systems: A Defeasible Argumentation Approach. *Energies* **2019**, *12*, 662. [CrossRef]
27. Zeng, M.; Yu, T.; Wang, X.; Su, V.; Nguyen, L.T.; Mengshoel, O.J. Improving Demand Prediction in Bike Sharing System by Learning Global Features. In Proceedings of the Machine Learning for Large Scale Transportation Systems (LSTS)@ KDD-16, San Francisco, CA, USA, 13 August 2016.
28. Kaltenbrunner, A.; Meza, R.; Grivolla, J.; Codina, J.; Banchs, R. Urban cycles and mobility patterns: Exploring and predicting trends in a bicycle-based public transport system. *Pervasive Mob. Comput.* **2010**, *6*, 455–466. [CrossRef]
29. Yoon, J.W.; Pinelli, F.; Calabrese, F. Cityride: A predictive bike sharing journey advisor. In Proceedings of the 2012 IEEE 13th International Conference on Mobile Data Management (MDM), Bengaluru, Karnataka, India, 23–26 July 2012; pp. 306–311.
30. Li, Y.; Zheng, Y.; Zhang, H.; Chen, L. Traffic prediction in a bike-sharing system. In Proceedings of the 23rd SIGSPATIAL International Conference on Advances in Geographic Information Systems, Seattle, WA, USA, 3–6 November 2015; ACM: New York, NY, USA, 2015; p. 33.
31. Chen, L.; Zhang, D.; Wang, L.; Yang, D.; Ma, X.; Li, S.; Wu, Z.; Pan, G.; Nguyen, T.M.T.; Jakubowicz, J. Dynamic cluster-based over-demand prediction in bike sharing systems. In Proceedings of the 2016 ACM International Joint Conference on Pervasive and Ubiquitous Computing, Heidelberg, Germany, 12–16 September 2016; ACM: New York, NY, USA, 2016; pp. 841–852.
32. Liu, J.; Li, Q.; Qu, M.; Chen, W.; Yang, J.; Xiong, H.; Zhong, H.; Fu, Y. Station site optimization in bike sharing systems. In Proceedings of the 2015 IEEE International Conference on Data Mining, Atlantic City, NJ, USA, 14–17 November 2015; IEEE: Piscataway, NJ, USA, 2015; pp. 883–888.
33. Lin, L.; He, Z.; Peeta, S. Predicting station-level hourly demand in a large-scale bike-sharing network: A graph convolutional neural network approach. *Transp. Res. Pt. C-Emerg. Technol.* **2018**, *97*, 258–276. [CrossRef]
34. Dell'Olio, L.; Ibeas, A.; Moura, J.L. Implementing bike-sharing systems. *Proc. Inst. Civ. Eng.-Munici. Eng.* 2011, 164, 89–101. [CrossRef]
35. Chen, Q.; Sun, T. A model for the layout of bike stations in public bike-sharing systems. *J. Adv. Transp.* **2015**, *49*, 884–900. [CrossRef]
36. Lin, J.R.; Yang, T.H. Strategic design of public bicycle sharing systems with service level constraints. *Transp. Res. Part E Logist. Transp. Rev.* **2011**, *47*, 284–294. [CrossRef]
37. Bargar, A.; Gupta, A.; Gupta, S.; Ma, D. Interactive visual analytics for multi-city bikeshare data analysis. In Proceedings of the 3rd International Workshop on Urban Computing (UrbComp 2014), New York, NY, USA, 24 August 2014; Volume 45.
38. Vogel, P.; Greiser, T.; Mattfeld, D.C. Understanding Bike-Sharing Systems using Data Mining: Exploring Activity Patterns. *Procedia-Soc. Behav. Sci.* **2011**, *20*, 514–523. [CrossRef]
39. Froehlich, J.E.; Neumann, J.; Oliver, N. Sensing and predicting the pulse of the city through shared bicycling. In Proceedings of the Twenty-First International Joint Conference on Artificial Intelligence, Pasadena, CA, USA, 14–17 July 2009.

40. Bao, J.; He, T.; Ruan, S.; Li, Y.; Zheng, Y. Planning bike lanes based on sharing-bikes' trajectories. In Proceedings of the 23rd ACM SIGKDD International Conference on Knowledge Discovery and Data Mining, Halifax, NS, Canada, 13–17 August 2017; ACM: New York, NY, USA, 2017; pp. 1377–1386.
41. He, T.; Bao, J.; Li, R.; Ruan, S.; Li, Y.; Tian, C.; Zheng, Y. *Detecting Vehicle Illegal Parking Events Using Sharing Bikes' Trajectories*; ACM: New York, NY, USA, 2018.
42. Pemantle, R. A survey of random processes with reinforcement. *Probab. Surv.* **2007**, *4*, 1–79. [CrossRef]
43. Huang, F.; Qiao, S.; Peng, J.; Guo, B. A Bimodal Gaussian Inhomogeneous Poisson Algorithm for Bike Number Prediction in a Bike-Sharing System. *IEEE Trans. Intell. Transp. Syst.* **2018**, 1–10. [CrossRef]
44. Pearson, K. The problem of the random walk. *Nature* **1905**, *72*, 342. [CrossRef]
45. Weiss, G.H.; Rubin, R.J. Random walks: theory and selected applications. *Adv. Chem. Phys.* **1982**, *52*, 363–505.
46. Scalas, E.; Gorenflo, R.; Mainardi, F. Fractional calculus and continuous-time finance. *Physica A* **2000**, *284*, 376–384. [CrossRef]
47. Cupertino, T.H.; Carneiro, M.G.; Zheng, Q.; Zhang, J.; Zhao, L. A scheme for high level data classification using random walk and network measures. *Expert Syst. Appl.* **2018**, *92*, 289–303. [CrossRef]
48. Pons, P.; Latapy, M. Computing Communities in Large Networks Using Random Walks. *J. Graph Algorithms Appl.* **2006**, *10*, 191–218. [CrossRef]
49. Backstrom, L.; Leskovec, J. Supervised random walks: predicting and recommending links in social networks. In Proceedings of the Fourth ACM International Conference on Web Search and Data Mining, Hong Kong, China, 9–12 February 2011; ACM: New York, NY, USA, 2011; pp. 635–644.
50. Shaheen, S.A.; Chan, N.D.; Gaynor, T. Casual carpooling in the San Francisco Bay Area: Understanding user characteristics, behaviors, and motivations. *Transp. Policy* **2016**, *51*, 165–173. [CrossRef]
51. Cervero, R.; Duncan, M. Walking, Bicycling, and Urban Landscapes: Evidence From the San Francisco Bay Area. *Am. J. Public Health* **2003**, *93*, 1478–1483. [CrossRef] [PubMed]

© 2019 by the authors. Licensee MDPI, Basel, Switzerland. This article is an open access article distributed under the terms and conditions of the Creative Commons Attribution (CC BY) license (http://creativecommons.org/licenses/by/4.0/).

MDPI
St. Alban-Anlage 66
4052 Basel
Switzerland
Tel. +41 61 683 77 34
Fax +41 61 302 89 18
www.mdpi.com

Energies Editorial Office
E-mail: energies@mdpi.com
www.mdpi.com/journal/energies

www.ingramcontent.com/pod-product-compliance
Lightning Source LLC
LaVergne TN
LVHW070920170726
843515LV00005B/825
* 9 7 8 3 0 3 9 3 6 7 6 0 3 *